T0181714

THE FRONTIERS COLLECTION

THE FRONTIERS COLLECTION

Series Editors
A.C. Elitzur   L. Mersini-Houghton   T. Padmanabhan   M. Schlosshauer
M.P. Silverman   J.A. Tuszynski   R. Vaas

The books in this collection are devoted to challenging and open problems at the forefront of modern science, including related philosophical debates. In contrast to typical research monographs, however, they strive to present their topics in a manner accessible also to scientifically literate non-specialists wishing to gain insight into the deeper implications and fascinating questions involved. Taken as a whole, the series reflects the need for a fundamental and interdisciplinary approach to modern science. Furthermore, it is intended to encourage active scientists in all areas to ponder over important and perhaps controversial issues beyond their own speciality. Extending from quantum physics and relativity to entropy, consciousness and complex systems—the Frontiers Collection will inspire readers to push back the frontiers of their own knowledge.

More information about this series at http://www.springer.com/series/5342

For a full list of published titles, please see back of book or springer.com/series/5342

Brigitte Falkenburg · Margaret Morrison
Editors

Why More Is Different

Philosophical Issues in Condensed Matter
Physics and Complex Systems

 Springer

Editors
Brigitte Falkenburg
Faculty of Human Sciences and Theology
TU Dortmund
Dortmund
Germany

Margaret Morrison
Trinity College
University of Toronto
Toronto, ON
Canada

ISSN 1612-3018   ISSN 2197-6619   (electronic)
THE FRONTIERS COLLECTION
ISBN 978-3-662-52309-4   ISBN 978-3-662-43911-1   (eBook)
DOI 10.1007/978-3-662-43911-1

Springer Heidelberg New York Dordrecht London

Printed on acid-free paper

Springer is part of Springer Science+Business Media (www.springer.com)

Contents

Part II   Emergence

6   Why Is More Different? 91
Margaret Morrison

7   Autonomy and Scales 115
Robert Batterman

Chapter 1
Introduction

Brigitte Falkenburg and Margaret Morrison

This volume on philosophical issues in the physics of condensed matter fills a crucial gap in the overall spectrum of philosophy of physics. Philosophers have generally focused on emergence in debates relating to the philosophy of mind, artificial life and other complex biological systems. Many physicists working in the field of condensed matter have significant interest in the philosophical problems of reduction and emergence that frequently characterise the complex systems they deal with. More than four decades after Philip W. Anderson's influential paper *More is Different* (Anderson 1972) and his well known exchange with Steven Weinberg in the 1990s on reduction/emergence, philosophers of physics have begun to appreciate the rich and varied issues that arise in the treatment of condensed matter phenomena. It is one of the few areas where physics and philosophy have a genuine overlap in terms of the questions that inform the debates about emergence. In an effort to clarify and extend those debates the present collection brings together some well-known philosophers working in the area with physicists who share their strong philosophical interests.

The traditional definition of emergence found in much of the philosophical literature characterizes it in the following way: A phenomenon is emergent if it cannot be reduced to, explained or predicted from its constituent parts. One of the things that distinguishes emergence in physics from more traditional accounts in philosophy of mind is that there is no question about the "physical" nature of the emergent phenomenon, unlike the nature of, for example, consciousness. Despite these differences the common thread in all characterizations of emergence is that it depends on a hierarchical view of the world; a hierarchy that is ordered in some fundamental way. This hierarchy of levels calls into question the role of reduction

B. Falkenburg (✉)
Faculty of Human Sciences and Theology, Department of Philosophy
and Political Science, TU Dortmund, D-44221 Dortmund, Germany
e-mail: brigitte.falkenburg@tu-dortmund.de

M. Morrison
Department of Philosophy, Trinity College, University of Toronto, Toronto,
ON M5S 1H8, Canada
e-mail: mmorris@chass.utoronto.ca

© Springer-Verlag Berlin Heidelberg 2015
B. Falkenburg and M. Morrison (eds.), *Why More Is Different*,
The Frontiers Collection, DOI 10.1007/978-3-662-43911-1_1

in relating these levels to each other and forces us to think about the relation of parts and wholes, explanation, and prediction in novel ways.

In discussing this notion of a "hierarchy of levels" it is important to point out that this is not necessarily equivalent to the well known fact that phenomena at different scales may obey different fundamental laws. For instance, while general relativity is required on the cosmological scale and quantum mechanics on the atomic, these differences do not involve emergent phenomena in the sense described above. If we characterise emergence simply in terms of some "appropriate level of explanation" most phenomena will qualify as emergent in one context or another. Emergence then becomes a notion that is defined in a relative way, one that ceases to have any real ontological significance. In true cases of emergence we have generic, stable behaviour that *cannot* be explained in terms of microphysical laws and properties. The force of "cannot" here refers not to ease of calculation but rather to the fact that the micro-physics fails to provide the foundation for a physical explanation of emergent behaviour/phenomena. Although the hierarchical structure is certainly present in these cases of emergence the ontological status of the part/whole relation is substantially different.

What this hierarchical view suggests is that the world is ordered in some fundamental way. Sciences like physics and neurophysiology constitute the ultimate place in the hierarchy because they deal with the basic constituents of the world—fundamental entities that are not further reducible. Psychology and other social sciences generally deal with entities at a less fundamental level, entities that are sometimes, although not always, characterised as emergent. While these entities may not be reducible to their lower level constituents they are nevertheless *ontologically* dependent on them. However, if one typically identifies explanation with reduction, a strategy common across the sciences, then this lack of reducibility will result in an accompanying lack of explanatory power. But, as we shall see from the various contributions to this volume, emergent phenomena such as superconductivity and superfluidity, to name a few, are also prevalent in physics. The significance of this is that these phenomena call into question the reliance on reduction as the ultimate form of explanation in physics and that everything can be understood in terms of its micro-constituents and the laws that govern them.

The contributions to the collection are organized in three parts: reduction, emergence, and the part-whole-relation, respectively. These three topics are intimately connected. The reduction of a whole to its parts is typical of explanation and the practices that characterise physics; novel phenomena typically emerge in complex compound systems; and emergence puts limitations on our ability to see reduction as a theoretical goal. In order to make these relations transparent, we start by clarifying the concepts of reduction and emergence. The first part of the book deals with general issues related to reduction, its scope, concepts, formal tools, and limitations. The second part focuses on the characteristic features of emergence and their relation to reduction in condensed matter physics. The third deals with specific models of the part-whole-relation used in characterizing condensed matter phenomena.

1.1 Reduction

Part I of the book embraces four very different approaches to the scope, concepts, and formal tools of reduction in physics. It also deals with the relation between reduction and explanation, as well as the way limitations of reduction are linked with emergence. The first three papers are written by condensed matter physicists whose contributions to the collection focus largely on reduction and its limitations. The fourth paper, written by a philosopher-physicist, provides a bridge between issues related to reduction in physics and more philosophically oriented approaches to the problem.

On the Success and Limitations of Reductionism in Physics by *Hildegard Meyer-Ortmanns* gives an overview of the scope of reductionist methods in physics and beyond. She points out that in these contexts ontological and theoretical reduction typically go together, explaining the phenomena in terms of interactions of smaller entities. Hence, for her, ontological and theoretical reduction are simply different aspects of methodological reduction which is the main task of physics; a task that aims at explanation via part-whole relations (ontological reduction) and the construction of theories describing the dynamics of the parts of a given whole (theoretical reduction). This concept of "methodological" reduction closely resembles what many scientists and philosophers call "mechanistic explanation" (see Chap. 13). The paper focuses on the underlying principles and formal tools of theoretical reduction and illustrates them with examples from different branches of physics. She shows how the same methods, in particular, the renormalization group approach, the "single step" approaches to pattern formation, and the formal tools of quantum field theory, are used in several distinct areas of research such as particle physics, cosmology, condensed matter physics, and biophysics. The limitations of methodological reduction in her sense are marked by the occurrence of strong emergence, i.e., non-local phenomena which arise from the local interactions of the parts of a complex system.

Barbara Drossel's contribution reminds us that the thorny problem of theoretical reduction in condensed matter physics deals, in fact, with three theories rather than two. *On the Relation between the Second Law of Thermodynamics and Classical and Quantum Mechanics* reviews the foundations of the thermodynamic arrow of time. Many physicists and philosophers take for granted that the law of the increase in entropy is derived from classical statistical mechanics and/or quantum mechanics. But how can irreversible processes be derived from reversible deterministic laws? Drossel argues that all attempts to obtain the second law of thermodynamics from classical mechanics include additional assumptions which are extraneous to the theory. She demonstrates that neither Boltzmann's H-theorem nor the coarse graining of phase-space provide a way out of this problem. In particular, coarse graining as a means for deriving the second law involves simply specifying the state of a system in terms of a finite number of bits. However, if we regard the concept of entropy as based on the number of possible microstates of a closed system, then this approach obviously begs the question. She emphasizes that quantum mechanics also fails to resolve the reduction problem. Although the Schrödinger equation justifies the assumption of a finite number of possible microstates, it does not explain the irreversibility and stochasticity of the second law.

Joachim Ankerhold addresses another complex reduction problem in the intersection of quantum mechanics, thermodynamics and classical physics, specifically, the question of how classical behaviour emerges from the interactions of a quantum system and the environment. The well-known answer is that the dissipation of the superposition terms into a thermal environment results in decoherence. *Dissipation in Quantum Mechanical Systems: Where is the System and Where is the Reservoir?* shows that issues surrounding this problem are not so simple. Given that the distinction between a quantum system and its environment is highly problematic, the concept of an open quantum system raises significant methodological problems related to ontological reduction. Condensed matter physics employs 'system + reservoir' models and derives a reduced density operator of the quantum system in order to describe decoherence and relaxation processes. The 'system + reservoir' picture depends on the epistemic distinction of the relevant system and its irrelevant surroundings. But, due to quantum entanglement it is impossible to separate the system and the reservoir, resulting in obvious limitations for the naïve picture. The paper shows that the model works only for very weak system-reservoir interactions based on a kind of perturbational approach; whereas in many other open quantum systems it is difficult to isolate any "reduced" system properties. However, due to a separation of time scales the appearance of a (quasi-) classical reduced system becomes possible, even in the deep quantum domain.

Rafaela Hillerbrand in her contribution entitled *Explanation via Microreduction: On the Role of Scale Separation for Quantitative Modelling* argues that scale separation provides the criterion for specifying the conditions under which ontological reduction can be coupled with theoretical or explanatory reduction. She begins by clarifying the philosophical concepts of reduction. The distinction between "ontological" and "explanatory" reduction employed here is based on the opposition of ontology and epistemology, or the distinction between what there is and what we know. Ontological reduction is "micro-reduction", similar to Meyer-Ortmann's concept of ontological reduction. Theoretical reduction is based on knowledge and can be further divided into "epistemic" reduction (tied to the DN- or deductive nomological model of explanation), and explanatory reduction in a broader sense, the main target of Hillerbrand's investigation. Her paper discusses scale separation and the role it plays in explaining the macro features of systems in terms of their micro constituents. She argues that scale separation is a necessary condition for the explanatory reduction of a whole to its parts and illustrates this claim with several examples (the solar system, the laser, the standard model of particle physics, and critical phenomena) and a counter-example (fluid dynamic turbulence). Her main conclusion is that micro-reduction with scale separation gives rise to a special class of reductionist models.

1.2 Emergence

The papers in Part II take a closer look at the limitations of reduction in order to clarify various philosophical aspects of the concept of emergence. According to the usual definition given above, emergent phenomena arise out of lower-level entities, but they cannot be reduced to, explained nor predicted from their micro-level base. Given that solids, fluids and gases consist of molecules, atoms, and subatomic particles, how do we identify emergent phenomena as somehow distinct from their constituents? What exactly is the relation between the micro- and the macro-level, or the parts and the emergent properties of the whole? A crucial concept is autonomy, that is, the independence of the emergent macro-properties. But the term "emergent" means that such properties are assumed to arise out of the properties and/or dynamics of the parts. How is this possible and what does this mean for the autonomy or independence of emergent phenomena?

Margaret Morrison focuses on the distinction of epistemic and ontological independence in characterizing emergence and how this is distinguished from explanatory and ontological reduction. *Why and How is More Different?* draws attention to the fact that the traditional definition of emergence noted above can be satisfied on purely epistemological grounds. However, taking account of Anderson's seminal paper we are presented with a notion of emergence that has a strong ontological dimension—that the whole is *different* from its parts. Since the phenomena of condensed matter physics are comprised of microphysical entities the challenge is to explain how this part/whole relation can be compatible with the existence of ontologically independent macro-properties; the properties we characterize as emergent. For example, all superconducting metals exhibit universal properties of infinite conductivity, flux quantization and the Meissner effect, regardless of the microstructure of the metal. However, we typically explain superconductivity in terms of the micro-ontology of Cooper pairing, so in what sense are the emergent properties independent/autonomous? Understanding this micro-macro relation is crucial for explicating a notion of emergence in physics. Morrison argues that neither supervenience nor quantum entanglement serve to explain the ontological autonomy of emergent phenomena. Nor can theoretical descriptions which involve approximation methods etc., explain the appearance of generic, universal behaviour that occurs in phase transitions. The paper attempts a resolution to the problem of ontological independence by highlighting the role of spontaneous symmetry breaking and renormalization group methods in the emergence of universal properties like infinite conductivity.

Robert Battermann's contribution entitled *Autonomy and Scales* also addresses the problem of autonomy in emergent behaviour but from a rather different perspective, one that has been ignored in the philosophical literature. He focuses on a set of issues involved in modelling systems across many orders of magnitude in spatial and temporal scales. In particular, he addresses the question of how one can explain and understand the relative autonomy and safety of models at continuum scales. He carefully illuminates why the typical battle line between reductive "bottom-up" modelling and 'top-down' modelling from phenomenological theories is overly simplistic. Understanding the philosophical foundations implicit in the physics of continuum scale problems requires a new type of modelling framework. Recently multi-scale models have been successful in showing how to upscale from statistical atomistic/molecular models to continuum/hydrodynamics models. Batterman examines these techniques as well as the consequences for our understanding of the debate between reductionism and emergence. He claims that there has been too much focus on what the actual fundamental level is and whether non-fundamental (idealized) models are dispensable. Moreover, this attention to the "fundamental" is simply misguided. Instead we should focus on proper modeling techniques that provide bridges across scales, methods that will facilitate a better understanding of the relative autonomy characteristic of the behavior of systems at large scales.

Paul Humphreys paper *'More is Different ... Sometimes'* presents a novel and intriguing interpretation of Philip Anderson's seminal paper 'More Is Different'. While Anderson's paper is explicit in its arguments for the failure of construction methods in some areas of physics, Humphreys claims that it is inexplicit about the consequences of those failures. He argues that as published, Anderson's position is obviously consistent with a reductionist position but, contrary to many causal claims, does not provide evidence for the existence of emergent phenomena. Humphreys defines various emergentist positions and examines some recent undecidability results about infinite and finite Ising lattices by Barahona and by Gu et al. He claims that the former do not provide evidence for the existence of ontologically emergent states in real systems but they do provide insight into prediction based accounts of emergence and the limits of certain theoretical representations. The latter results bear primarily on claims of weak emergence and provide support for Anderson's views. Part of the overall problem, Humphreys argues, is that one should not move from conclusions about the failure of constructivism and undecidability to conclusions about emergence without an explicit account of what counts as an entity being emergent and why. The failure of constructivism in a particular instance is not sufficient for emergence in the sense that the inability in practice or in principle to compute values of a property is insufficient for the property itself to count as emergent. He leaves as an open question the pressing problem of determining what counts as a novel physical property.

Continuing with the attempt to clarify exactly what it at stake in the characterization of emergent phenomena, *Sorin Bangu's* paper *Neither Weak, Nor Strong? Emergence and Functional Reduction* draws attention to the long history behind the clarification of the concept of emergence, especially in the literature on the metaphysics of science. Notions such as 'irreducibility', 'novelty' and 'unpredictability' have all been invoked in an attempt to better circumscribe this notoriously elusive idea. While Bangu's paper joins that effort, it also contributes a completely different perspective on the clarificatory exercise. He carefully examines a class of familiar physical processes such as boiling and freezing, processes generically called 'phase transitions' that are characteristic of what most philosophers and physicists take to be paradigm cases of emergent phenomena. Although he is broadly sympathetic to some aspects of the traditional characterization, the paper questions what kind of emergence these processes are thought to instantiate. Bangu raises this issue because he ultimately wants to depart from the orthodoxy by claiming that the two types of emergence currently identified in the literature, 'weak' and 'strong', do not adequately characterize the cases of boiling and freezing. The motivation for his conclusion comes from an application of Kim's (1998, 1999, 2006) 'functional' reduction model (F-model). When applied to these cases one finds that their conceptual location is undecided with respect to their 'emergent' features. As it turns out, their status depends on how one understands the idealization relation between the theories describing the macro-level (classical thermodynamics) and the micro-level (statistical mechanics) reality.

1.3 Parts and Wholes

Part III consists of four papers that focus on the part-whole-relation in order to shed light on the methods, successes and limitations of ontological reduction in condensed matter physics and beyond. The first two contributions discuss the explanatory power of the many-body systems of condensed matter physics but with a very different focus in each case. The last two papers investigate the dynamical aspects of the part-whole relation and their ontological consequences. Today, ontological reduction is often characterised in terms of "mechanistic explanation". A mechanism typically consists of some type of causal machinery according to which the properties of a whole are caused by the dynamic activities of the parts of a compound system. In that sense the papers in this section of the book deal, broadly speaking, with the successes and limitations of mechanistic explanation, even though the term is only used specifically in Kuhlman's paper.

Andreas Hüttemann, Reimer Kühn, and Orestis Terzidis address the question of whether there is an explanation for the fact that, as Fodor put it, the micro-level "converges on stable macro-level properties", and whether there are lessons from this explanation for similar types of issues. *Stability, Emergence and Part-Whole-Reduction* presents an argument that stability in large (but non-infinite) systems can be understood in terms of statistical limit theorems. They begin with a small simulation study of a magnetic system that is meant to serve as a reminder of the fact that an increase of the system size leads to reduced fluctuations in macroscopic properties. Such a system exhibits a clear trend towards increasing stability of macroscopic (magnetic) order and, as a consequence, the appearance of ergodicity breaking, i.e. the absence of transitions between phases with distinct macroscopic properties in finite time. They describe the mathematical foundation of the observed regularities in the form of limit theorems of mathematical statistics for independent variables (Jona-Lasinio 1975) which relates limit theorems with key features of large scale descriptions of these systems. Generalizing to coarse-grained descriptions of systems of interacting particle systems leads naturally to the incorporation of renormalization group ideas. However, in this case Hüttemann et al. are mainly interested in conclusions the RNG approach allows one to draw about system behaviour away from criticality. Hence, an important feature of the analysis is the role played by the finite size of actual systems in their argument. Finally, they discuss to what extent an explanation of stability is a reductive explanation. Specifically they claim to have shown that the reductionist picture, according to which the constituents' properties and states determine the behaviour of the compound system, and the macro-phenomena can be explained in terms of the properties and states of the constituents, is neither undermined by stable phenomena in general nor by universal phenomena in particular.

Axel Gelfert's contribution *Between Rigor and Reality: Many-Body Models in Condensed Matter Physics* focusses on three theoretical dimensions of many-body models and their uses in condensed matter physics: their structure, construction, and confirmation. Many-body models are among the most important theoretical 'workhorses' in condensed matter physics. The reason for this is that much of condensed matter physics aims to explain the macroscopic behaviour of systems consisting of a large number of strongly interacting particles, yet the complexity of this task requires that physicists turn to simplified (partial) representations of what goes on at the microscopic level. As Gelfert points out, because of the dual role of many-body models as models of physical systems (with specific physical phenomena as their explananda) as well as mathematical structures, they form an important sub-class of scientific models. As such they can enable us to draw general conclusions about the function and functioning of models in science, as well as to gain specific insight into the challenge of modelling complex systems of correlated particles in condensed matter physics. Gelfert's analysis places many-body models in the context of the general philosophical debate about scientific models (especially the influential 'models as mediators' view), with special attention to their status as mathematical models. His discussion of historical examples of these models provides the foundation for a distinction between different strategies of model construction in condensed matter physics. By contrasting many-body models with phenomenological models, Gelfert shows that the construction of many-body models can proceed either from theoretical 'first principles' (sometimes called the ab initio approach) or may be the result of a more constructive application of the formalism of many-body operators. This formalism-based approach leads to novel theoretical contributions by the models themselves (one example of which are so-called 'rigorous results'), which in turn gives rise to cross-model support between models of different origins. A particularly interesting feature of Gelfert's deft analysis is how these different features allow for exploratory uses of models in the service of fostering model-based understanding. Gelfert concludes his paper with an appraisal of many-body models as a specific way of investigating condensed matter phenomena, one that steers a middle path 'between rigor and reality'.

Brigitte Falkenburg investigates the ontological status of quasi-particles that emerge in solids. Her paper *How Do Quasi-Particles Exist?* shows that structures which emerge within a whole may, in fact, be like the parts of that whole, even though they seem to be higher-level entities. Falkenburg argues that quasi-particles are real, collective effects in a solid; they have the same kinds of physical properties and obey the same conservation laws and sum rules as the subatomic particles that constitute the solid. Hence, they are ontologically on a par the electrons and atomic nuclei. Her paper challenges the philosophical view that quasi-particles are fake entities rather than physical particles and counters Ian Hacking's reality criterion: "If you can spray them, they exist". Because of the way quasi-particles can be used as markers etc. in crystals, arguments against their reality tend to miss the point. How, indeed, could something that contributes to the energy, charge etc. of a solid in accordance with the conservation laws and sum rules be classified as "unreal"? In order to spell out the exact way in which quasi-particles exist, the paper discusses their particle properties in extensive detail. They are compared in certain respects to those of subatomic matter constituents such as quarks and the virtual field quanta of a quantum field. Falkenburg concludes that quasi-particles are ontologically on par with the real field quanta of a quantum field; hence, they are as real or unreal as electrons, protons, quarks, photons, or other quantum particles. Her contribution nicely shows that the questions of scientific realism cannot be settled without taking into account the emergent phenomena of condensed matter physics, especially the conservation laws and sum rules that connect the parts and whole in a hierarchical view of the physical world.

Meinard Kuhlmann's paper addresses the important issue of mechanistic explanations which are often seen as the foundation for what is deemed explanatory in many scientific fields. Kuhlmann points out that whether or not mechanistic explanations are (or can be) given does not depend on the science or the basic theory one is dealing with but rather on the type of object or system (or 'object system') under study and the specific explanatory target. As a result we can have mechanistic and non-mechanistic explanations in both classical and quantum mechanics. *A Mechanistic Reading of Quantum Laser Theory* shows how the latter is possible. Kuhlmann's argument presents a novel approach in that quantum laser theory typically proceeds in a way that seems at variance with the mechanistic model of explanation. In a manner common in the treatment of complex systems, the detailed behaviour of the component parts plays a surprisingly subordinate role. In particular, the so-called "enslaving principle" seems to defy a mechanistic reading. Moreover, being quantum objects the "parts" of a laser are neither located nor are they describable as separate entities. What Kuhlmann shows is that despite these apparent obstacles, quantum laser theory provides a good example of a mechanistic explanation in a quantum-physical setting. But, in order to satisfy this condition one needs to broaden the notion of a mechanism. Although it is tempting to conclude that these adjustments are ad hoc and question-begging, Kuhlmann expertly lays out in detail both how and why the reformulation is far more natural and less drastic than one may expect. He shows that the basic equations as well as the methods for their solution can be closely matched with mechanistic ideas at every stage. In the quantum theory of laser radiation we have a decomposition into components with clearly defined properties that interact in specific ways, dynamically producing an organization that gives rise to the macroscopic behavior we want to explain. He concludes the analysis by showing that the structural similarities between semi-classical and quantum laser theory also support a mechanistic reading of the latter.

Most of the contributions to this volume were presented as talks in a workshop of the Philosophy Group of the German Physical Society (DPG) at the general spring meeting of the DPG in Berlin, March 2012. Additional papers were commissioned later. We would like to thank the DPG for supporting the conference from which the present volume emerged, and Springer for their interest in the publication project and for allowing us the opportunity to put together a volume that reflects new directions in philosophy of physics. A very special thank you goes to Angela Lahee from Springer, who guided the project from the initial proposal through to completion. In addition to her usual duties she wisely prevented us from giving the volume the amusing but perhaps misleading title "Condensed Metaphysics" (as an abbreviation of "The Metaphysics of Condensed Matter Physics"). Not only did she offer many helpful suggestions for the title and the organisation of the book, but showed tremendous patience with the usual and sometimes unusual delays of such an edition. Finally we would like to thank each of the authors for their contributions as well as their willingness to revise and reorganise their papers in an effort to make the volume a novel and we hope valuable addition to the literature on emergence in physics.

Part I
Reduction

Chapter 2
On the Success and Limitations of Reductionism in Physics

Hildegard Meyer-Ortmanns

2.1 Introduction

Natural sciences, and in particular physics, can look back over a track record of increasing predictive power with regard to the outcome of time evolutions, control, as well as the design of experiments of far-reaching technological and practical importance. But, their success has also brought deeper insights into the underlying laws that govern a wide variety of phenomena. Without doubt this success is based on methodological reductionism, i.e., the attempt to reduce explanations to smaller constituents (although not necessarily the smallest) and to explain phenomena completely in terms of interactions between fundamental entities. Included in the scope of methodological reductionism is theoretical reductionism, wherein one theory with limited predictive power can be obtained as a limiting case of another theory, just as Newtonian mechanics is included in general relativity. From the beginning we should emphasize that reductionism does not preclude emergent phenomena. It allows one to predict some types of emergent phenomena, as we shall see later, even if these phenomena are not in any sense the sum of the processes from which they emerge.

In the following, emergence is understood as involving new, sometimes novel properties of a whole that are not shared by its isolated parts. Emergent phenomena generated this way are therefore intrinsically nonlocal. Within the reductionistic approach we understand them as a result of local interactions, as characteristic of approaches in physics. Emergent phenomena definitely extend beyond simple formation of patterns, such as those in mass and pigment densities. Functionality may be an emergent property as well, as in cases where systems are built up of cells, the fundamental units of life. In our later examples, we shall not refer to "weak emergence", where a phenomenon is predicted as a result of a model.

H. Meyer-Ortmanns (✉)
Jacobs University, Campus Ring 8, 28759 Bremen, Germany
e-mail: h.ortmanns@jacobs-university.de

© Springer-Verlag Berlin Heidelberg 2015
B. Falkenburg and M. Morrison (eds.), *Why More Is Different*,
The Frontiers Collection, DOI 10.1007/978-3-662-43911-1_2

Instead, we shall usually mean "strong emergence", where nonlocal phenomena arise from local interactions.

Emergent features are not restricted to patterns in an otherwise homogeneous background. "Being alive" is also an emergent property, arising from the cell as the fundamental unit of life. The very notion of complexity is a challenging one. In our context, systems are considered genuinely complex if they show behavior that cannot be understood by considering small subsystems separately. Our claim is a modest one—it is not that complex systems can be understood in all their facets by analyzing them locally, but that complexity can often be reduced by identifying local interactions. Moreover, we do not adopt the extreme view which considers complex systems as inherently irreducible, thereby requiring a holistic approach. The art is to focus on just those complex features that can be reduced and broken up into parts. Why this is not a fruitless endeavor is the topic of the Sect. 2.2.

Section 2.2 deals with the "recipes" responsible for the success. They are abstract guiding principles as well as the use of symmetries, such as the principle of relativity and Lorentz covariance, leading to the theory of special relativity; the equivalence principle and covariance under general coordinate transformations, leading to the theory of general relativity, as well as the gauge principle and invariance under local gauge transformations (complemented by the Higgs mechanism for the electroweak part), leading to the standard model of elementary particle physics. These theories have extraordinary predictive power for phenomena that are governed by the four fundamental interactions; three of them involve the realm of subatomic and atomic physics at one end of the spatial scale, while gravity becomes the only relevant interaction on cosmic scales, where it determines the evolution of the universe.

Interactions on macro or intermediate mesoscopic scales, like the nano and microscales, are in principle produced by the fundamental interactions when composite objects are formed. In practice, they can be derived using a phenomenological approach that involves models valid on this particular scale. Beyond the very formulation of these models, reductionism becomes relevant as soon as one tries to bridge the scales, tracing phenomena on the macroscale back to those on the underlying scales. "Tracing back" means predicting changes on the macro and mesoscopic scales produced by changes on the microscale. A computational framework for performing these bridging steps is the renormalization group approach of Kogut and Wilson (1974), Wilson (1975) and Kadanoff (1977). The framework of the renormalization group goes far beyond critical phenomena, magnetism, and spin systems (see Sect. 2.2.2.1).[1] More generally, but very much in the spirit of the renormalization group, we now have what is called multiscale analysis, with applications in a variety of different realms. In general, it involves links between different subsystems, with each subsequent system having fewer degrees of freedom than its predecessor. The new system may still be

---

[1] For further applications, see also Meyer-Ortmanns and Reisz (2007).

complex, but the iterative nature of the procedure gradually reduces the complexity (see Sect. 2.2.2.4 below).

Sometimes one is in the fortunate situation where no intermediate steps are needed to bridge the scales from micro to macro behaviour. This can happen when static spatial patterns form on large scales according to rules obeyed by the constituents on the smaller scale, or when shock waves propagate over large distances and transport local changes. We shall illustrate pattern formation with applications as different as galaxy formation in the universe as well as spots and stripes on animals in the realm of living systems. We shall further use dynamical pattern formation in evolving strains of bacteria to illustrate increasing mathematical complexity, as more and more features are simultaneously taken into account. This leads us to conclude that any candidate for an equation of "everything" will be constrained to describe only "something", but not the whole (see Sect. 2.2.4).

One may wonder why there is in general a need for bridging the scales in intermediate steps. Why not use a single step by exploiting modern computer facilities? After all, it is now possible to simulate a virus in terms of its atomic constituents (an example will be sketched in Sect. 2.2.5). The very same example we use to illustrate the power of up-to-date computer simulations could in principle also serve to demonstrate typical limitations of reductionism. Reductionism, pushed to its extreme, makes the description clumsy. It does not identify the main driving mechanisms on intermediate scales that underlie the results on larger scales. Reductionism then falls short of providing explanations in terms of simple mechanisms, which is what we are after. A more serious worry is that new aspects, properties, features, and interpretations may emerge on the new scale that a computer experiment may inevitably miss. In a fictive dialogue we debate the positions of an extreme reductionism with a more moderate version. As an example of the moderate version, we consider DNA from the perspective of physics and computer science. Even if there are no equations of theories that deserve the attribute "of everything", or if a multitude of disciplines must be maintained in the future, one may still wonder whether some further steps towards a universal theory of complex systems are possible. Such steps will be sketched in Sect. 2.4.

2.2 On the Success of Reductionism

2.2.1 Symmetries and Other Guiding Principles

Physical theories are primarily grounded in experiment in that they are proposed to reproduce and predict experimental outcomes. What distinguishes them from optimized fits of data sets is their range of applicability and their predictive power. Some of these theories deserve to be classified as fundamental. To this class belongs the theories of special, general relativity and the standard model of particle physics.

In this section we would like to review some guiding principles that led to their construction, restricting what would otherwise be a multitude of models to a limited set.

2.2.1.1 The Special Relativity Principle

According to Albert Einstein the Special Relativity Principle postulates that all inertial frames are totally equivalent for the performance of all physical experiments, not only mechanical ones, but also electrodynamics. (In this way, Einstein was able to eliminate absolute space as the carrier of light waves and electromagnetic fields.) Insisting in particular on the constancy of the velocity of light propagation in all inertial frames, one is then led in a few steps to the conclusion that the coordinates of two inertial frames must be related by Lorentz transformations. (First one can show that the transformations must be linear, then one can reduce considerations to special transformations in one space direction, and finally one shows that the well-known γ-factor takes its familiar form $\gamma = 1/\sqrt{1 - v^2/c^2}$.)[2] If a physical law is invariant under these special Lorentz transformations, and also under spatial rotations and translations in space and time, it holds in any inertial system. The corresponding transformations between two arbitrary inertial systems are then Poincaré transformations. The reductionism arising from the special relativity principle (including the constancy of the velocity of light) leads to the restriction to formulate laws in inertial frames in flat space as equations between tensors under Poincaré transformations. In particular, it restricts the choice of Lagrangians, such as the Lagrangian of electrodynamics, to scalars under these transformations.

2.2.1.2 The Equivalence Principle and General Relativity

Einstein wanted to eliminate "absolute space" in its role in distinguishing inertial frames as those in which the laws take a particularly simple form. He put the equivalence principle at the center of his considerations. According to the (so-called) weak equivalence principle, inertial and gravitational mass are proportional for *all* particles, so that all particles experience the same acceleration in a given gravitational field. This suggests absorbing gravity into geometry, the geometry of spacetime, to which all matter is exposed. The equivalence principle led Einstein to formulate his general relativity theory. From Newton's theory, it was already known that mechanics will obey the same laws in a freely falling elevator as in a laboratory that is not accelerated and far away from all attracting masses. Einstein extrapolated this fact to hold, not only for the laws of mechanics, but so that all local, freely falling, nonrotating labs are fully equivalent for the performance of *all* experiments.

---

[2] For the derivation see, for example, Rindler (1969).

(Therefore the simple laws from inertial systems now hold everywhere in space, but only locally, so that special relativity also becomes a theory that is supposed to hold only locally.) This extrapolation amounts to the postulate that the equations in curved space-time should be formulated as tensor equations under general coordinate transformations, where curved space-time absorbs the effect of gravity. Due to the homogeneous transformation behavior of tensors, the validity of tensor equations in one frame ensures their validity in another frame, related by general coordinate transformations. This postulate finally led Einstein to the theory of general relativity that has been confirmed experimentally to a high degree of accuracy.

2.2.1.3 Gauge Theories of the Fundamental Interactions

In the previous sections on the relativity principle, the postulated symmetries referred to transformations of the space-time coordinates and restricted the form of physical laws. In the theories of strong, weak, and electromagnetic interactions, we have to deal with internal symmetries. Here it is not only the right choice of the symmetry group which is suggested by conserved matter currents, but also the prescription of how to implement the dynamics of matter and gauge fields that lead to the construction of gauge theories and finally to the standard model of particle physics.

Hermann Weyl was the first to consider electromagnetism as a local gauge theory of the $U(1)$ symmetry group (Weyl 1922). Let us first summarize the steps in common to the construction of electromagnetic, strong, and weak interactions as a kind of "recipe". As result of Noether's theorem, one can assign a global (space-independent) continuous symmetry to a conserved matter current. The postulate of local gauge invariance then states that the combined theory of matter and gauge fields should be invariant under local (that is space-dependent) gauge transformations. Obviously, a mass term, which is bilinear in the matter fields $\psi, \bar{\psi}$ and contains a partial derivative, violates this invariance. To compensate for the term that is generated from the derivative of the space dependent phase factors in the gauge transformations, one introduces a so-called minimal coupling between the matter fields and the gauge fields, replacing the partial derivative by the covariant derivative in such a way that the current is covariantly conserved. It remains to equip the gauge fields with their own dynamics and construct the gauge field strengths in such a way that the resulting Lagrangian is invariant under local gauge transformations.

Let us demonstrate these steps in some more detail. Under local gauge transformations, matter fields $\psi_c(x)$ transform according to

$$\psi_c(x) \rightarrow (\exp(-i\Theta^a(x)T_a))_{cc'}\psi_{c'}(x) \equiv (g(x)\psi)_c(x). \tag{2.1}$$

Here T_a are the infinitesimal generators of the symmetry group $SU(n)$ in the fundamental representation, $\Theta^a(x)$ are the space dependent group parameters,

$a = 1, \ldots, N$, with N the dimension of the group, and $c, c' = 1, \ldots, n$, where n characterizes the symmetry group $SU(n)$. Gauge fields $\mathcal{A}_{\mu,cc'}(x) = \bar{g} A_\mu^a(x)(T_a)_{cc'}$, which are linear combinations of the generators T_a, with $\bar{g}$ a coupling constant, transform inhomogeneously according to

$$\mathcal{A}'_\mu(x) = g(x)(\mathcal{A}_\mu - i\partial_\mu)g^{-1}(x). \qquad (2.2)$$

In general, this equation should be read as a matrix equation. In the language of differential geometry, the gauge field corresponds to a connection, it allows one to define a parallel transport of charged vector fields $\psi_c(x)$ from one space-time point x, along a path $\mathcal{C}$ to another point y. This parallel transport can then be used to compare vector fields from different space-time points in one and the same local coordinate system. It thus leads to the definition of the covariant derivative D_μ:

$$[(\partial_\mu + i\mathcal{A}_\mu)\psi(x)]_c dx^\mu =: (D_\mu\psi)_c(x)dx^\mu. \qquad (2.3)$$

The path dependence of the parallel transport is described infinitesimally by the field strength tensor $\mathcal{F}_{\mu\nu}(x)$ with $\mathcal{F}_{\mu\nu}(x) = \bar{g} F_{\mu\nu}^a(x)T_a$. In terms of gauge fields, $F_{\mu\nu}^a(x)$ is given by

$$F_{\mu\nu}^a(x) = \partial_\mu A_\nu^a(x) - \partial_\nu A_\mu^a(x) - \bar{g}f^{abc}A_\mu^b(x)A_\nu^c(x), \qquad (2.4)$$

with structure constants f^{abc} specific to the gauge group. For $U(1)$, the last term vanishes, whence it is characteristic of the nonabelian gauge groups. In geometric terms, the field strength tensor is given by the commutator of the covariant derivatives

$$D_\mu D_\nu - D_\nu D_\mu = i\mathcal{F}_{\mu\nu}(x). \qquad (2.5)$$

The last equation reflects the fact that the parallel transport is path dependent if there is a non-vanishing field strength, in very much the same way as the parallel transport of a vector in Riemannian space depends on the path if the space is curved. The field strength of the gauge fields then transforms under local gauge transformations $g(x)$ according to the adjoint representation of the symmetry group:

$$\mathcal{F}_{\mu\nu}(x) \rightarrow g(x)\mathcal{F}_{\mu\nu}(x)g^{-1}(x). \qquad (2.6)$$

This construction principle leads for (quantum) electrodynamics to the familiar Lagrange density

$$\mathcal{L} = -\frac{1}{4}F_{\mu\nu}F^{\mu\nu} + \bar{\psi}^{(l)}(x)(i\gamma^\mu D_\mu - M_l)\psi^{(l)}(x), \qquad (2.7)$$

with $D_\mu = \partial_\mu - ieA_\mu$. By construction it is invariant under the local $U(1)$ transformations given by

$$
\begin{aligned}
A_\mu &\rightarrow A_\mu(x) + \partial_\mu \Theta(x), \\
\psi^{(l)}(x) &\rightarrow e^{ie\Theta(x)} \psi^{(l)}(x), \\
\bar{\psi}^{(l)}(x) &\rightarrow \bar{\psi}^{(l)}(x) e^{-ie\Theta(x)},
\end{aligned}
\tag{2.8}
$$

where $\Theta(x)$ is a space-dependent phase, l labels the electron or muon, ψ is a Dirac spinor representing the matter fields, and A_μ represents the photons. For quantum chromodynamics, the resulting Lagrange density takes the same form as in (2.7):

$$
\mathcal{L} = -\frac{1}{4} F^a_{\mu\nu} F^{a,\mu\nu} + \bar{\psi}(x)(i\gamma^\mu D_\mu - M)\psi(x),
\tag{2.9}
$$

where we have suppressed the indices of the mass matrix M and the quark fields ψ. Note that ψ here carries a multi-index α, f, c, where α is a Dirac index, f a flavor index, and c a color index, and all indices are summed over in $\mathcal{L}$. The gauge transformations (2.1) can be specialized to $T_a = \frac{1}{2}\lambda_a$, with $a = 1 \dots, 8$ and λ_a the eight Gell-Mann matrices, and $c, c' = 1, 2, 3$, for the three colors of the $SU(3)$ color symmetry. The covariant derivative takes the form $D_\mu = \partial_\mu - ig\frac{\lambda_a}{2}A^a_\mu(x)$, where the gauge fields A^a_μ now represent the gluon fields mediating the strong interaction, and the field strength tensor $F^a_{\mu\nu}$ is given by (2.4) with structure constants f^{abc} from $SU(3)$. Note that the quadratic term in (2.4) represents the physical fact that gluons are also self-interacting, in contrast to photons. So in spite of the same form of (2.7) and (2.9), the physics thereby represented is as different as are quantum electrodynamics and quantum chromodynamics.

Finally, the combined action of electromagnetic and weak interactions is constructed along the same lines, with an additional term in the action that implements the Higgs mechanism, to realize the spontaneous symmetry breaking of $SU(2)_w \times U(1)$ to $U(1)_e$ (where the subscript w stands for "weak" and e for "electromagnetic") and give masses to the vector bosons W^+, W^- and Z mediating the weak interactions.

The similarities between the local gauge theories and general relativity become manifest in the language of differential geometry and point to the deeper reasoning behind what we called initially a "recipe" for how to proceed. In summary, the pendants in local gauge theories and general relativity are the following:

- The local space $\mathcal{H}(x)$ of charged fields $\psi(x)$ with unitary structure corresponds to the tangential space with local metric $g_{\mu\nu}(x)$ and Lorentz frames.
- The local gauge transformations correspond to general coordinate transformations.
- The gauge fields $\mathcal{A}_{\mu,cc'}(x)$, defining the connection in the parallel transport, correspond to the Christoffel symbols, which describe the parallel transport of tangential vectors on a Riemannian manifold.

- The covariant derivatives correspond to each other; from an abstract point of view, the idea behind their construction and their transformation behavior is the same.
- The field strength tensor $\mathcal{F}_{\mu\nu,cc'}(x)$ corresponds to the Riemann curvature tensor R^i_{kmn}.

The overarching mathematical structure between gauge theories and general relativity is formulated in the theory of fiber bundles.

The standard model of particle physics has been confirmed experimentally to a very high level of accuracy. It brings order into the otherwise confusing zoo of elementary particles. Before its formulation in terms of local gauge theories, there were a variety of effective, phenomenological models, based merely upon symmetry requirements, with a rather limited range of predictions as compared to the standard model. It was the recognition of electrodynamics as a local gauge theory of $U(1)$, and the extension of the postulate of local gauge invariance to the strong and weak interactions, together with an implementation along the lines of differential geometry and general relativity, that led to its construction and eventual success. Seminal contributions along the way were made by Weyl (1922), Yang and Mills (1954), Glashow (1961), Weinberg (1967), Salam (1968), to name but a few,[3] and Kibble (1961), Sciama (1962), Hehl et al. (1976), and others in relation to the formulation of gravitational theory as a local gauge theory.

2.2.2 Bridging the Scales from Micro to Macro

2.2.2.1 The Renormalization Group Approach

The renormalization group is neither a group nor a universal procedure for calculating a set of renormalized parameters from a set of starting values. It is a generic framework with very different realizations. Common to them is the idea of deriving a set of new (renormalized) parameters, characteristic of a larger scale, from a first set of parameters, characteristic of the underlying smaller scale, while keeping some long-distance physics unchanged. The degrees of freedom are partitioned into disjoint subsets. Specific to the renormalization group is a partitioning according to length scale, or equivalently, according to high and low momentum modes. These successive integrations of modes according to their momentum or length scales are the result of an application of the renormalization group equations. Since the change in scale goes along with a reduction in the number of degrees of freedom, the iterated procedure should lead to a simpler description of the system of interest, which is

---

[3] For a reference on "The Dawning of Gauge Theory", see also the book (O'Raifeartaigh 1997) with the same title.

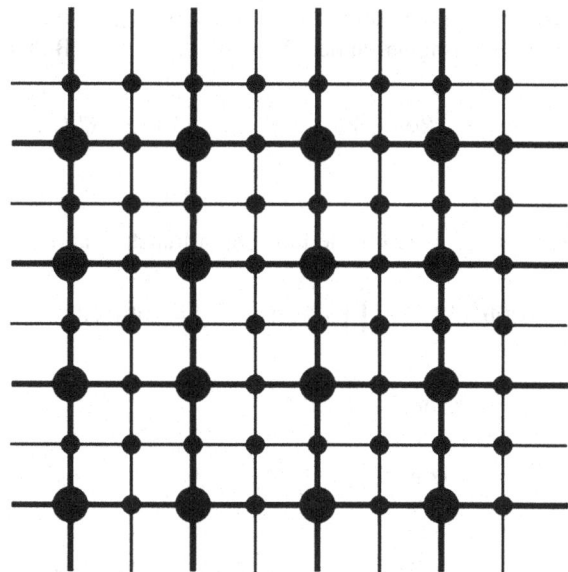

Fig. 2.1 Square lattice of size 8×8 in $D = 2$, with an assigned block lattice of scale factor $l = 2$

often the long-distance physics. The renormalization group provides a computational tool. There are different ways to implement the change of scale but for simplicity, we choose the framework of block spin transformations as an example.

2.2.2.2 Renormalization Group for a Scalar Field Theory

Let us consider the theory of a single (spin-zero bosonic) scalar field on a D-dimensional hypercubic lattice $\Lambda = (a\mathbb{Z})^D$ with lattice spacing a. In order to describe the block spin transformations, we have to introduce some definitions. The scalar fields make up a vector space $\mathcal{F}_\Lambda$ of real-valued fields $\Phi : \Lambda \to \mathbb{R}$. The action S_Λ is a functional on these fields, and the partition function is given by $Z_\Lambda(J)$ with

$$Z_\Lambda(J) = \int \mathcal{D}\mu_\Lambda(\Phi) \exp\left(-S_\Lambda(\Phi) + (J, \Phi)_\Lambda\right), \tag{2.10}$$

where $J \in \mathcal{F}_\Lambda$ stands for an external current and $\mathcal{D}\mu_\Lambda(\Phi) = \prod_{x\in\Lambda} d\Phi(x)$, while $(J, \Phi)_\Lambda = a^D \sum_{x\in\Lambda} J(x)\Phi(x)$. Let us now define a block lattice Λ_l for $l \in \mathbb{N}$ by decomposing Λ into disjoint blocks. Each block consists of l^D sites of Λ (see Fig. 2.1 for $D = 2$ and $l = 2$).

The renormalization transformation R_l of the action S_Λ, leading to the effective action S'_Λ, defined on the original lattice Λ, is defined via the Boltzmann factor

$$\exp\left(-S'_\Lambda(\Psi)\right) \equiv \exp\left(-(R_l S_\Lambda)(\Psi)\right) = \int \mathcal{D}\mu_\Lambda(\Phi) P(\Phi, s_l\Psi) \exp\left(-S_\Lambda(\Phi)\right),$$

$$(2.11)$$

where a commonly used choice for the block spin transformation is given by

$$P(\Phi, s_l\Psi) = \prod_{x\in\Lambda_l} \delta[(s_l\Psi)(x) - \frac{1}{b}(C\Phi)(x)],$$

$$(2.12)$$

with block averaging operator C defined by

$$(C\Phi)(x) := \frac{1}{l^D} \sum_{y\in\text{block}(x)} \Phi(y).$$

$$(2.13)$$

So the effective action $S'_\Lambda(\Psi)$ in terms of the block variables Ψ results from a path integral over all Φ of the Boltzmann factor $e^{-S_\Lambda(\Phi)}$ under the constraint that the average value of $\Phi(y)$ over all sites of a block (normalized over the number of these sites, viz., l^D, and multiplied by a scale factor $1/b$) takes a prescribed value $\Psi(x)$ for each block, where the block is labeled by $x \in \Lambda_l$. The rescaling operation $(s_l\Psi)(x) := \Psi(x/l)$ with $x \in \Lambda_l$ accounts for the fact that lengths and distances on the block lattice reduce by a factor of $1/l$ when measured in units of the block lattice distance as compared to units on the original lattice. Note that the way the constrained integration is realized here amounts to an integration over short wavelength fluctuations with a wavelength λ satisfying $a < \lambda < la$. The effective action S'_Λ then describes the fluctuations of the scalar field with wavelength $\lambda > la$. The choice of the block variable as some rescaled average (by a factor of b) over the variables of the block is plausible as long as the average values are good representatives for the whole ensemble of variables. If the variables are elements of a certain group like $SU(3)$, the sum is no longer an $SU(3)$ element, so it is obvious that the naive averaging procedure will not always work. For block variables that are spins with two possible values, one may use the majority rule instead. If the majority of spins points "up" within a block, the representative is chosen to point up, etc.

2.2.2.3 Renormalization Group for the Ising Model

An alternative option for selecting a block variable is decimation. In the simplest case of the Ising model in one dimension, decimation amounts to the choice of block spins as spins of a subset of the original chain, for example, choosing as block spin the spins of every second site. In the partition function, this amounts to taking the partial trace. When the resulting partition function in terms of a trace over the reduced set of variables is cast in the same form as the original one with the Ising action in the Boltzmann factor, one can read off what the renormalized parameters (of the effective Ising action on the coarse scale) are in terms of the original parameters (of the Ising action on the underlying scale). Writing for the Ising action

$$S = -k \sum_{<ij>} s_i s_j - h \sum_i s_i - \sum_i c, \qquad (2.14)$$

with coupling k, external field h, and constant c,[4] $s_i \in \{\pm 1\}$ and $<ij>$ denoting nearest neighbors, one obtains the following relations for the renormalized parameters k', h', c' in terms of those of the original action:

$$\begin{aligned}
\exp(2h') &= \exp(2h) \cosh(2k+h) / \cosh(2k-h), \\
\exp(4k') &= \cosh(2k+h) \, \cosh(2k-h) / \cosh^2 h, \\
\exp(4c') &= \exp(8c) \cosh(2k+h) \, \cosh(2k-h) \, \cosh^2 h.
\end{aligned} \qquad (2.15)$$

For the derivation of these relations, see for instance (Yeomans 1992). Here it should be emphasized that the result is exact, since the effective action on twice the scale, resulting from the decimation, can be exactly cast into the form of an Ising model without truncation of other terms. Usually, further terms are generated in the effective action under a renormalization group transformation. Exact self-similarity with respect to the action is the exception rather than the rule.

The successes of the renormalization group in relation to critical phenomena are as follows:

• One can explain why second-order phase transitions fall into so-called universality classes.
• One can predict an upper critical dimension for a given universality class.
• One can derive scaling relations as equalities between different critical exponents.
• Once can explain why critical exponents take the same values if calculated from above or below the critical point.

Our examples of a scalar field theory and an Ising model are simple dynamical systems. In the vicinity of critical points, the conjecture of self-similar actions over the spatial scales could be suggested, for example, by visualizing the blocks of aligned spins: at the critical point the linear block size varies overall length scales, so it is natural to define the block spin in such a way as to represent a whole block by another spin. In general, block variables should be representative of the whole block, in the sense that they project onto the appropriate degrees of freedom. They should also guarantee that the dynamics becomes simple again on the coarse scale in terms of these variables. A choice of block variables for which the effective

---

[4] Keeping the constant c from the beginning, although it appears to be redundant in (2.14), one can claim that the new action has the same form as the old one.

action contained many more terms than the original one, of which all remained relevant under iteration of the procedure, would fail. Therefore there is a certain skill involved in making an appropriate choice, and this is an act of "cognition" that cannot be automated in a computer simulation.

Although from a rigorous point of view it may be quite hard to control the truncations of terms in the effective actions, a closer look at the mesoscopic or macroscopic scales reveals that there do exist new sets of variables that afford a relatively simple phenomenological description, even if it is not feasible to derive them within the renormalization group approach. In a similar spirit to the renormalization group is multi-scale modeling, which we consider next.

2.2.2.4 Multi-scale Modeling

Multi-scale modeling is an approach that is now also used outside physics, in engineering, meteorology and computer science, and in particular in materials science. There it can be used to describe hierarchically organized materials like wood, bones, or membranes (Fratzl and Weinkammer 2007). Typically, one has a few levels with non-self-similar dynamics and different variables for each level. The output of one level serves as input for the next. The number of variables for each level should be limited and tractable. As an example, let us indicate typical levels in bones. Starting with a cube of the order of a few mm^3 of trabecular structure of a human vertebra, we zoom to the microscale of a single vertebra, then further to the sub-microscale of a single lamella, then to the nanoscale of collagen fibers, and finally to the genetic level (see, for example, Wang and Gupta 2011). Conversely (and different research groups may proceed either top-down or bottom-up), once we succeed in understanding the regulation of bones on the genetic level and its impact on intracellular processes, identifying its impact on cell–cell communication, then on ensembles of cells, and finally on the whole organism, we will be able to treat bone diseases on the genetic level. Moreover, by understanding self-healing and restructuring processes of bones, we may be able to imitate nature's sophisticated design of bone material. Here reductionism leads to the "industry" of bionics and biomimetics, which is already booming for many applications.

2.2.3 When a Single Step Is Sufficient: Pattern Formation in Mass and Pigment Densities

Sometimes one is in the lucky situation that the scales from micro to macro distances can be bridged in a "single step", that is in a single set of equations, as the inherent local rules lead to patterns on a coarse, macroscopic scale. We would like to give two examples from very different areas, cosmology and biology.

2.2.3.1 Pattern Formation in the Universe

Let us first illustrate the great success of reductionism for the example of galaxy formation in the universe. For a detailed description we refer to (Bartelmann 2011) and references therein. Based on two symmetry assumptions (that of homogeneity and isotropy of space) and the theory of general relativity, one first derives the Friedmann equations. Together with the first law of thermodynamics and an equation of state for matter, one arrives at the standard model for the structure and evolution of the universe. In particular, assuming that dark matter gives the main contribution to the total mass and that it can be approximated as pressureless, the equations governing the evolution of the dark matter density are the continuity equation for mass conservation, the Euler equation for momentum conservation, and the gravitational field equation of Newtonian physics (here the Einstein equations of general relativity are not even needed in view of the final accuracy). The three equations can be combined into one, viz.,

$$\ddot{\delta} + 2H\dot{\delta} - 4\pi G\bar{\rho}\delta \; = \; 4\pi G\bar{\rho}\delta^2 + \frac{1}{a^2}\nabla\delta\nabla\Phi + \frac{1}{a^2}\partial_i\partial_j[(1+\delta)u_iu_j]. \qquad (2.16)$$

Here $\delta \equiv (\rho - \bar{\rho})/\bar{\rho}$ are the density fluctuations around the mean density $\bar{\rho}$, H denotes the Hubble function, G the gravitational constant, Φ the gravitational potential in Newtonian gravity, $-\nabla\Phi$ the gravitational force, $\vec{u}$ the velocity of matter with respect to the mean Hubble expansion of the universe, and $a(t)$ the scale factor entering the Friedmann model. By deriving the initial density fluctuations in the early universe from the observed CMB (cosmic microwave background) data under the assumption of cold dark matter and evolving the resulting Gaussian fluctuations with respect to (2.16) in time, one can reproduce the formation first of filamentary or sheet-like structures as they are experimentally observed in large-scale galaxy surveys, then of galaxy clusters and galaxies. The quantities to be compared between experiment and theory are the power spectra of the variance of fluctuation amplitudes, and these are measurable over a vast range of scales. This is clearly a striking success of the reductionistic approach, starting from symmetries and basic laws of physics, to arrive at an equation for the mass density fluctuations that is able to reproduce structure formation from the scale of about 1 Mpc (1 megaparsec $\approx 3 \times 10^{22}$ m)[5] to cosmic scales.

On the other hand one has to admit that the considered observable, the mass density fluctuations, is a universal but simple characteristic that keeps its meaning over a vast range of scales, and the only relevant force on large scales is gravity. The formation of functional structures of the kind occurring within biological units is incomparably more difficult to trace back to a few "ingredients" as input, as the

---

[5] This estimate depends on the applicability of (2.16). One megaparsec is supposed to be an estimate for a lower bound if the mass density fluctuations refer to dark matter. Density fluctuations of gas in cosmic structures may be described as a fluid down to even smaller scales.

struggle to construct precursors of biological cells so clearly demonstrates in the study of artificial life.

2.2.3.2 Pattern Formation in Animal Coats

We would like to add another example of pattern formation, based on different mechanisms and in a very different range of applications. This is pattern formation in animal coats. The mechanism goes back to Turing (1952) who suggested that, under certain conditions, chemicals can react and diffuse in such a way as to produce steady state spatial patterns of chemical or morphogen concentrations. For two chemicals $A(\vec{r}, t)$, $B(\vec{r}, t)$, the reaction–diffusion equations take the form

$$\frac{\partial A}{\partial t} = F(A, B) + D_A \nabla^2 A,$$
$$\frac{\partial B}{\partial t} = G(A, B) + D_B \nabla^2 B, \tag{2.17}$$

where F and G are nonlinear functions of A and B that determine the reaction kinetics and D_A, D_B are the diffusion constants. According to Turing's idea, spatially inhomogeneous patterns can evolve by diffusion driven instability if $D_A \neq D_B$. (Special cases of $F(A, B)$ and $G(A, B)$ include the activator–inhibitor mechanism, suggested by Gierer and Meinhardt (1972).) In particular, one can derive the necessary conditions on the reaction kinetics and the diffusion coefficients for a process of reacting and diffusing morphogens, once the set of differential equations (2.17) is complemented by an appropriate set of initial and boundary conditions. Murray suggested (Murray 1980, 1981) that a single (reaction–diffusion) mechanism could be responsible for the versatile patterns in animal coats. It should be noticed that pattern formation does not directly refer to the pigment density. What matters are the conditions on the embryo's surface at the time of pattern activation. Pattern formation first refers to morphogen prepatterns for the animal coat markings, and it requires a further assumption that subsequent differentiation of the cells to produce melanin simply reflects the spatial pattern of morphogen concentration. Solving the differential equations for parameters and geometries which are adapted to those of animals (like the surface of a tapering cylinder to simulate patterns forming on tails) leads to remarkable agreement between general and specific features of mammalian coat patterns.

The important role that the size and shape of the domain have on the final pattern (spots and stripes and the like) can be tested through a very different realization of the related spatial eigenvalue problem

$$\nabla^2 \vec{W} + K^2 \vec{W} = 0. \tag{2.18}$$

In relation to animal coats, $\vec{W}$ represents the fluctuations about the steady state concentration in dimensionless units, and solutions reflect the initial stages of pattern formation. In a different realization of (2.18), $\vec{W}$ represents the amplitude of vibrations of a membrane, a thin plate, or a drum surface, since their vibrational modes also solve (2.18). Time-average holographic interferograms on a plate, excited by sound waves, nicely visualize patterns and their dependence on the size and form of the plate. Varying the size (for a plate, this is equivalent to varying the forcing frequency) generates a sequence of patterns in the interferograms that bear a striking resemblance to a sequence of simulated animal coats of varying sizes (Murray 1993 and references therein).

The reductionism here amounts to explaining the variety of patterns in animal coats in terms of a single set of equations (2.17) with specified parameters and functions F and G, assuming that the morphogen prepatterns get transferred to the finally observed pigment patterns. Moreover, analysis of this set of equations also allows one to study the sensitivity to initial conditions. In the biological implementation, it is the initial conditions in the embryonic stage that matter for the final pattern. It should be noticed that beyond the universal characteristics of these patterns (spots, stripes), the precise initial conditions at the time of activation of the reaction–diffusion mechanism determine the individuality of the patterns. Such individuality is important for kin and group recognition among animals. The role the pattern plays in survival differs between different species. If it does play a role, the time of activation in the embryonic phase should be well controlled. These remarks may give some hints regarding the fact that, although the basic mechanism behind pattern formation sounds rather simple, its robust implementation in a biological context raises a number of challenging questions for future research.

2.2.4 From Ordinary Differential Equations to the Formalism of Quantum Field Theory: On Increasing Complexity in the Description of Dynamic Strains of Bacteria

Dynamic strains of bacteria provide another example of pattern formation. In this section we use these systems to demonstrate a generic feature that is observed whenever one increases the number of basic processes that should be included in one and the same description at the same time. It is not only, and not necessarily, the number of variables or the number of equations that increases along with the different processes, but also the mathematical complexity of the required mathematical framework. Including different processes in one and the same framework should be contrasted with treating them separately, in certain limiting cases. These processes may be their self-reproduction and destruction, or birth and death events, caused by their mutual interactions, all this leading to a finite lifetime of individuals and therefore to demographic noise, and also their movement via diffusion or active motion, and their assignment to a spatial grid, restricting the range of individual interactions.

As an example of such a system that can be treated in various limiting cases, we consider strains of bacteria in microbial experiments in a Petri dish, reproducing, diffusing, going extinct, and repressing or supporting each other according to certain rules. All these features can be observed in the spatially extended May–Leonard model (May and Leonard 1975). This model is intended to describe generic features of ecosystems such as contest competition, realized via selection, and scramble competition, realized via reproduction. The selection events follow the cyclic rock–paper–scissors game according to the following rules:

$$AB \rightarrow A\emptyset,$$
$$BC \rightarrow B\emptyset, \qquad (2.19)$$
$$CA \rightarrow C\emptyset.$$

This means that the N individuals occur in three species A, B, C, where A consumes B at rate σ if they are assigned to neighboring sites on a two-dimensional grid of linear size L, and similarly B consumes C and C consumes A at rates that here are chosen to be the same for simplicity. The reproduction rules are:

$$A\emptyset \rightarrow AA,$$
$$B\emptyset \rightarrow BB, \qquad (2.20)$$
$$C\emptyset \rightarrow CC.$$

Hence, A reproduces at rate μ if the neighboring site of A is empty, and B and C accordingly. It is assumed that the lattice sites have a finite carrying capacity, viz., zero or one. In addition, the individuals are allowed to move. One option is to let the individuals exchange their position with a nearest neighbor at rate ϵ, leading to effective diffusion. Here we use a realization of the model as described in (Frey 2010). The overall goal is to predict the space-time evolution at large times as a function of the inherent parameters, and in particular to predict and characterize the kind of pattern formation that happens on the spatial grid (The conditions that ensure the coexistence of different species on the grid are of primary interest, in view of one of the core questions of ecology: the maintenance of biodiversity.).

If we want to approach the problem in full generality, as it has just been posed, we would need either numerical simulations or the quantum field theoretic framework[6] from the outset (for an example, Ramond (1989).) Instead, let us start with the various limiting cases and assume three species with a total of N individuals, interacting according to the rules of (2.19), (2.20) in all the following cases:

---

[6] The formalism of quantum field theory can be applied to systems which are fully classical. See (Mobilia et al. 2007)

1. N infinite, no spatial assignment, no explicit mobility.
   When the population size of the three species A, B, C goes to infinity, and the individuals are well mixed in the sense that they are neither assigned to the space continuum nor to a spatial grid, we obtain the following set of ordinary (nonlinear) differential equations (ODEs) for the corresponding concentrations of the three species a, b, c:

$$\begin{aligned}
\partial_t a &= a[\mu(1 - \rho) - \sigma c], \\
\partial_t b &= b[\mu(1 - \rho) - \sigma a], \\
\partial_t c &= c[\mu(1 - \rho) - \sigma b].
\end{aligned} \tag{2.21}$$

Here ρ denotes the overall density, and μ and σ are as defined below (2.19) and (2.20), respectively. These equations deterministically predict the time evolution of species concentrations.

2. N finite fluctuating, no spatial assignment, no explicit mobility.
   Let us keep the species well mixed, not assigned to a grid, but keeping N finite and fluctuating. Now the appropriate description is in terms of a master equation for the probability P to find N_i individuals of species i ($i \in \{A, B, C\}$) at time t (under the assumption of a Markov process):

$$\partial_t P(N_i, t) = f(P(N_i, t), P(N_i \pm 1, t); \mu, \sigma), \tag{2.22}$$

where the right-hand side is a function f, depending on the probabilities for finding states with N_i or $N_i \pm 1$ individuals at time t, the latter being states from which a decay or creation of one individual contributes to a change in $P(N_i, t)$. Note that we now obtain a deterministic description of the probabilities of finding a certain configuration of species rather than of the concentrations themselves.

3. N infinite, concentrations spatially assigned as $\vec{a}(\vec{r})$, with diffusion.
   Now we obtain a deterministic reaction–diffusion equation, that is, a set of coupled partial differential equations (PDEs) of the form

$$\partial_t \vec{a}(\vec{r}, t) = D\Delta\vec{a} + \vec{F}(\vec{a}), \tag{2.23}$$

where $\vec{a}(\vec{r}, t)$ denotes the vector of three space-time dependent concentrations of species, and $\vec{F}$ the appropriate function of $\vec{a}$, given by the right-hand side of (2.21), while D is the diffusion constant with $D = \epsilon/2N$ finite for $N \to \infty$, so that ϵ has to increase accordingly.

4. N finite fluctuating, species spatially assigned to a grid, but high mobility.
   Becoming more realistic and keeping N finite and fluctuating while moving in space, it is in the low-noise approximation that the spatiotemporal evolution of the system can be described by concentrations of the species, evolving in the

space-time continuum, and the result can be cast in a set of stochastic partial differential equations (SPDEs):

$$\partial_t \vec{a}(\vec{r}, t) = D \triangle \vec{a}(\vec{r}, t) + \vec{F}(\vec{a}) + C(\vec{a})\vec{\xi}. \tag{2.24}$$

Here $\xi_i(\vec{r}, t)$, $i = A, B, C$ denotes Gaussian white noise, $\triangle$ is the Laplacian, and $\vec{F}(\vec{a})$ is the former reaction term. In principle, a noise term in these equations could have three origins: the stochasticity of chemical reactions according to (2.19) and (2.20), a finite fluctuating number N when it is not forced to be conserved, and the motion of individuals. The noise term $C(\vec{a})\vec{\xi}$ in (2.24) represents only the noise in the reactions (2.19) and (2.20) (along with non-conserved N), where the noise amplitudes $C(\vec{a})$ are sensitive to the system's configurations $\vec{a}(\vec{r}, t)$. As argued in (Frey 2010), noise due to mobility can be neglected as compared to the other sources in this limit (4). Note that it is only in the low-noise approximation that one obtains equations for the concentrations $\vec{a}$ rather than for the species numbers N_i, and assigned to a space continuum rather than to a grid. The effect of finite N is indirectly represented by the noise term. Equivalent to (2.24) would be the corresponding Fokker–Planck equations for the respective concentration probabilities.

5. N finite fluctuating, species spatially assigned to a grid, and low mobility.
   This no longer corresponds to a limiting case. When in contrast to case (4) the exchange rate of species is no longer high compared with the reaction events, the former continuum description in terms of SPDEs breaks down, the low-noise approximation fails, and the field theoretic formalism is required as an analytical complement to numerical simulations. One should express the transition amplitude of an initial to a final occupation number distribution as a path integral over all occupation number configurations, where the path is weighted by an appropriate action that should be derived from the corresponding master equation. It depends on the specific interaction rules and the model parameters. The essential assumption is that each configuration is uniquely characterized by the occupation numbers N_i of the lattice site $\vec{r}$ with species $i = A, B, C$.

In summary, the subsequent inclusion of demographic fluctuations (due to annihilation, creation, local interactions), spatial organization, and diffusion leads to increasing complexity in the required mathematical description. The actual solutions to the equations of our example can be found in Frey (2010) and references therein. In the detailed version of Frey (2010), it is interesting to focus on the qualitative changes in the predictions that are missed by projecting on certain limits like high mobility or infinite population size. For example, in the limit discussed under (1), transitions in which certain species go extinct would be completely missed.

In this example, our agents were bacteria, but it is obvious that the bacteria may be replaced by more or less complex agents like humans or chemical substances, while adapting the wording accordingly. The principal need for simultaneously including all these aspects into a single framework comes from the requirement of not missing those phenomena that only occur in the simultaneous presence of all the available options when the system evolves. In general, it is not only that predictions obtained in certain limiting cases may be modified outside the validity range of the limit, but, more importantly, additional phenomena may also show up.

These results also shed some light on the meaning of any "Equation of Everything". As soon as "everything" is cast into the form of a (differential) equation, it corresponds to a projection. A theory, cast into the form of a path integral, itself corresponds to a selection of cases in which the system can be described by the time evolution of functionals, depending on configurations in terms of discrete occupation numbers. This framework is indeed quite generic, though apparently not the most abstract one that is achievable (see Sect. 2.4).

2.2.5 Large-Scale Computer Simulations: A Virus in Terms of Its Atomic Constituents

As we have seen in the previous sections, bridging the scales requires a number of iterations between the micro and macro levels, unless the mechanism of pattern formation is the main focus of interest. In the iterative procedure, on each level one has to deal with a number of degrees of freedom that is considerably reduced as compared to the original full set on the smallest scale. Most important is therefore a suitable choice of variables on the intermediate scales, in terms of which the dynamics becomes tractable. On the other hand, in view of today's high level of computer power one may wonder about using a brute force method instead, and numerically simulating the laws from the micro- to the macroscale in a single step. This means following the many paths (worldlines) of all individual constituents over a certain amount of time from the subatomic or atomic scale to the scale of macromolecules, deriving biophysics in terms of particle physics via large-scale computer simulations. Indeed, simulating a nano-machine like a virus in terms of its atomic constituents is feasible, as the following example of the mosaic tobacco virus will show.

This virus is rod-shaped. Its ribonucleic acid (RNA) is surrounded by a coat of proteins. Its name comes from the fact that it causes mosaic-like symptoms in plants. The virus causes abnormal cellular function that does not kill the plant but stunts growth. Its deleterious effect is not restricted to the tobacco plant, since it can infect other plants as well. In order to combat the virus and control its spreading, one should understand its key regulators and survival mechanisms. A step in this direction was taken by the molecular dynamics simulations described in Freddolino et al. (2006) with up to one million atoms over a time interval of 50 ns. The simulated virus consists of a capsid, composed of 60 identical copies of a single protein, and a 1,058 kb RNA genome. It is modeled by 949 nucleotides out of the complete genome, arranged into 30 double-stranded helical segments of 9 base pairs each. Only the RNA backbone in the 30 stems was resolved at atomic resolution. These numbers should give some hints on the structure of the virus, but they are not sufficient for a complete description, which cannot be provided in this context. One of the results states that the capsid becomes unstable without RNA. This has implications for assembly and infection mechanisms.

The reductionism here amounts to the fact that one "only" has to know Newton's equations of motion (as an approximation of the original quantum mechanical system) with the appropriate forces and the appropriate interaction topology on the atomic level to let the system evolve. The success of this approach lies in the important insights it provides on how to control this virus. A further advantage of such in silico experiments is the ease of manipulating the virus. One may rarefy certain constituents more easily than in vitro, to check their effect on the evolution of the remainder. In this way the simulation can explore strategies to combat the virus that can be used later in vivo. However, it should not be thought that this kind of realization of reductionism, down to the atomic level, comes cheaply. There is no such thing as a free lunch, and this is a case in point. Rather advanced computer algorithms are needed along with parallel computing and a sophisticated network between the parallel processors to obtain the results in a reasonable time (see Freddolino et al. (2006) for further details). On today's laptops, these simulations would take decades. In general, the CPU time of molecular dynamics simulations is easily of the order of millions of CPU-hours.

Still we may be tempted to extrapolate the computer power of today and ask when we shall be able to simulate *humans*. If we were able to do so, would that mean that biophysics explains life? Apart from numerous medical and technical applications, let us list some typical topics addressed in biophysics, such as the efficiency of nano-machines, the control of cell logistics,[7] the communication of nerve cells, the emergence of macroscopic features from local forces acting at junctions of the cytoskeleton, and the intermediate energy states of cellular fusion and fission events. Addressing these topics using tools from biophysics will lead to deeper insights into the fascinating nano and micro worlds, leaving open, however, the core mystery of how life emerges from non-living ingredients, the topical goal of artificial life studies.

We would like to add a remark on the demands and feasibility of computer simulations today. In the example of the tobacco virus, the computer simulations bridge the scales from atoms to macromolecules. Therefore to outsiders of particle physics, it may come as a surprise how demanding first-principle calculations are when they merely bridge the subnuclear scale of quarks and gluons to the nuclear scale of mesons and baryons. Let us consider typical first-principle calculations in (lattice) quantum chromodynamics, implemented on a space-time grid of size $128^3 \times 256$. In order to calculate decay constants or excited states of the mass spectrum of mesons or baryons to an accuracy of the order of a few percent (for the experts, using the framework of staggered fermions), the number of required floating point operations is estimated to take of the order of 200 years if 10^{12} floating

---

[7] It is instructive to consider cells as factories with a production output that has to be delivered at a certain time.

point operations are executed per second (in short, 200 Tflops/s-years). This means that about 50,000 processing units (cores) would be required to run for 1 year (Khaleel 2009). This should indicate the effort required for numerical first-principle calculations, even if one stays within the realm of particle physics. The compact notation for the Lagrangian in (2.9), the one that enters the path integral formulation of quantum chromodynamics, may obscure the extent of this effort to the outsider. The fundamental interactions between elementary particles are anything but elementary to handle.

Whatever an appropriate formulation of this kind of suggested "conservation law" might be, we may say that, whenever reductionism is pushed to its extreme, it will exact a price from us. It is as if hidden subtleties pop up and take revenge for the decomposition into simple constituents, so that the effort required of us is kept constant.

2.3 Limitations of Reductionism

2.3.1 A Fictive Dialogue For and Against Extreme Reductionism

To present the different mind-sets with respect to an extreme version of reductionism, we start with a fictive dialogue between two representatives, one, called PRO, extrapolating the power of reductionism to the extent that particle physics ultimately explains everything, and CON, pointing out the limitations of reductionism.

PRO: Knowing the laws on the fundamental scales, that is, on the scale of elementary particle physics, we can explain the whole world, at least in principle.

CON: Certainly some aspects, such as the reason why atoms stay together unless the temperature gets sufficiently high, and things like that, but not all. You are confusing the fundamental scale with fundamental phenomena. What is special about the scale of particle physics, or, more precisely, about the Planck scale (as the smallest scale we can talk about using classical notions of space), is its extreme value, but fundamental phenomena and fundamentally new phenomena occur also on larger scales.

PRO: I disagree, for the following reason. Let me be overly optimistic and extrapolate the present computer power by orders of magnitude as compared to today. We do know the laws of the four fundamental interactions in their ultimate form of local gauge theories (including gravity). Let us postulate initial conditions, which are extrapolated backwards in time from the CMB data observed today, to represent the soup of elementary particles at the end of the Planck epoch and the beginning of a universe in which we may use classical notions of space and time. Now let us rerun the tape of evolution in silico, at least in some sequences, with a supercomputer of a future generation. My claim is that the supercomputer would reproduce the formation of protons and neutrons out of the quarks and gluons, nuclei, atoms, molecules, macromolecules, proteins, nanomachines, and to cut a long sequence short, life. Or, to quote M.L. Goldberger and W. Panofsky from Anderson (2011): "Other branches of physics [than particle physics]...are interesting, challenging, and of great importance. The objectives are not a search for fundamental laws as such, these having been known...since the 1920s. Rather, they are the application of these laws."

CON: To directly reply with another quote from Anderson (2011): "If broken symmetries, localization, fractals and strange attractors are not "fundamental", what are they?" Even if computer simulations were to predict the emergence of molecular machines in the same way as they predict the emergence of spontaneous symmetry breaking in a magnet, they would fail to explain emergent features on each scale in simple terms, they would fail to provide a coarse-grained description on the coarse scale, to generate acts of cognition like interpreting DNA as a carrier of information.

PRO: On each scale there are new emergent features due to the act of cognition, I agree, but cognition is subjective. It is only ourselves who interpret DNA as a carrier of information, and even the selection of objects that are declared to be elementary on the coarse scale is down to us; it is our choice. This cannot and need not be reproduced by the supercomputer when it starts from the initial set of elementary particles and evolves them towards the nanoworld.

CON: Cognition of the kind declaring that composed objects are new elementary ones on a larger scale may sound subjective, but it has an objective pendant. Consider the fact that physical processes on the coarse scale may not be able to resolve the composition of composite objects. In processes with a typical momentum transfer that is much smaller than the mass scale, characteristic of the structure of the bound state, the structure cannot be resolved and the bound state will appear to be an ordinary particle. Fundamental fields and composite fields should then be treated on a completely equal footing (Ellwanger et al. 1994). As an example, consider mesons. On energy scales typical for mesons, which are quark–antiquark bound states, the particles are much more simply described in terms of mesonic degrees of freedom than in terms of their constitutive quark and gluonic degrees of freedom. Beyond the phenomenological level, here the mathematical derivation itself suggests to introduce new degrees of freedom on the lower energy scale. Last, but not least, changing the scale gives rise not only to new objects that should be considered as elementary, but also new interactions between them.

PRO: Why do you need your coarse-grained description, which is supposed to hold only on a single scale, the scale under consideration, if the supercomputer evolves the formation of larger and larger composed objects out of the fundamental ones? Let it go on and simulate humans.

CON: I need it precisely for my understanding in terms of simple mechanisms. I need this kind of understanding to abstract universal features from different realizations, to be inspired to new ideas, and in particular to design the new computer generations you are looking forward to.

2.3.2 DNA from the Standpoint of Physics
and Computer Science

Let us illustrate the CON perspective from the previous dialogue using the example of DNA. From the physics standpoint, DNA is a flexible macromolecule with a certain charge density and electrophoretic mobility, and a helix–coil transition. It stretches, melts, and changes twist under tension (Gross et al. 2011). It consists of two entangled polymers of helicoidal structure and its configuration depends on the degree of hydration and the ionic strength of the solution. The structure and state of DNA is also triggered and determined by its interaction with many other biomolecules, especially DNA-binding proteins or enzymes, which can compact, align, or bend DNA. Essentially two type of forces stabilize the double helix structure: hydrogen bonding between complementary bases and stacking interactions of the base pair plateaux. DNA is certainly a challenging object from the perspective of physics. During transcription and replication, it undergoes large conformational changes, like other polyelectrolytes. Due to large amplitude motions, nonlinear dynamics must be taken into account (Yakusheich 1998) and thermal fluctuations play an important role in its functioning. Earlier, physical experiments on DNA were performed by methods from condensed matter physics, while more recently, new powerful techniques have been developed. In particular, single-molecule experiments nowadays allow one to measure forces in the pN range (10^{-12} N), so that the elastic properties and interaction forces of single DNA molecules can be investigated directly (Bustamente et al. 2003). Together with experiments, theoretical models were developed which require tools from statistical physics and nonlinear dynamics. So it is definitely the laws of physics that both enable and constrain the basic functions and modes of performance of DNA.

However, no computer simulation in terms of its atomic constituents could create the insight that DNA acts as a carrier of genetic information, nor would it be able to establish a possible bridge to computer science that we want to sketch in the following. In an "act of cognition", one may wonder why DNA makes use of four characters (ACTG) in its alphabet and 20 amino acids plus one stopper. A look at searching problems in computer science may shed some light on this very choice of numbers (Patel 2001), although it is currently too early to claim that computer science will eventually explain this choice. A particular class of search problems there refers to search in an unsorted database with N distinct randomly arranged objects. The task is to locate a certain object by asking a set of Q questions. The set should be minimal. If the search is performed as a quantum search procedure, it has been shown by Grover (1997) that, for given N, the number of queries Q is determined by

$$(2Q + 1) \sin^{-1}\left(\frac{1}{\sqrt{N}}\right) = \frac{\pi}{2}, \tag{2.25}$$

so that asymptotically $Q = \pi\sqrt{N}/4$. Note that for given N, Q may not be an integer, so that a small error remains if it is chosen as such, but the search is accelerated by a factor of $\sqrt{N}$ as compared to classical search procedures. For small N, we obtain $Q = 1$ for $N = 4$ and $Q = 3$ for $N = 20.2$. As pointed out by Patel (2001) in this connection, base pairing during DNA replication corresponds to one yes/no query, to distinguish between the $N = 4$ possibilities of characters ACTG. The triplet code of DNA has $Q = 3$ consecutive nucleotide bases carrying $N = 21$ signals. Three base pairings between the t-RNA and m-RNA transfer this code to the amino acid chain. This may suggest that DNA provides the hardware that is best suited for the implementation of a kind of quantum search algorithm. A quantum search may be realized in classical hardware (Bhattacharya et al. 2002). Therefore the fact that DNA behaves as a whole as an object from classical physics is not an immediate objection.

This example shows that more and more phenomena emerge on the scales of biophysics, along with new phenomena, new properties of composites, and new rules governing their interactions. The qualitative variety proliferates so greatly that other disciplines than physics are needed to understand them, i.e., approaches from a range of different perspectives are needed.

2.4 Outlook: A Step Towards a Universal Theory of Complex Systems

In the previous sections we have pointed out that it is not meaningful to talk about a "World Formula" or a "Theory of Everything", although such extrapolations occur repeatedly in the literature. Still there are striking universal features across the scales and between dynamical systems. Does the mathematical framework developed so far capture them in a satisfactory way or can we go a step beyond stochastic partial differential equations and the field theoretic formalism, to a more abstract and more generic level of description?

Let us first summarize some striking facts, using the terminology of Mack (2001) in terms of "death", "growth", "motion", and "cognition". So far we have seen similar fundamental processes on very different scales such as annihilation in particle physics, deletion in chemical processes, extinction of species, which may be summarized under "death"; creation in particle physics, replication in biology, composition, fragmentation (as the reverse process), and recombination, summarized under "growth"; diffusion, drift, migration, summarized under "motion"; and, last but not least, the emergence of new degrees of freedom from scale to scale, the creation of new links between objects with matching internal structure, summarized under "cognition". Moreover, what induces death and growth in the former sense shares common features. We distinguish interactions as attractive or repulsive, activating or repressive, excitatory or inhibitory, short- or long-ranged, leading to binding and unbinding events, fusion or fission, selection or coexistence, competition or frustration (The very notion of frustration, familiar from the context of spin physics, is already a more generic concept than is usually assumed (see Mack (1981)).).

One may suspect that, although these analogies are evident, they hold at best on a superficial level. However, there is a formal manifestation in a framework that captures these analogies. It is the framework of (quantum) field theory, which can be applied whenever the state of the system can be described by a distribution of occupation numbers assigned to sites of a grid, and in which the time evolution refers to functionals of these occupation number distributions, as we mentioned earlier. However, when field theory is successfully applied to ecological systems[8], for example, this does not imply that biology is "physics" after all. It is merely the mathematical complexity of the description that is shared between the different areas of application. Still, although the language used in the field theoretic formalism is very generic, it may not be the most abstract one that is accessible for applications in physics.

A further step may be the framework of local category theory, as proposed by Mack (2001). Laws then correspond to regularities of relations, graphically represented as links, that is, relations between objects and agents, the latter being represented as nodes of a network. *Systems* are defined as such networks equipped with a minimal set of axioms. The axioms are from category theory, the mathematical theory of relations, extended by a notion of locality that is not restricted to locality in space-time. The dynamics of these defined systems is described in terms of certain *mechanisms*: motion, growth, death, and cognition, examples of which we gave above. Formally, mechanisms are conditional actions of basic local structural transformations. Motion, for example, is then characterized by the fact that indirect links become direct. The dynamics can be stochastic or deterministic. The only information put into the systems is *structure*. Systems differ by their structure, which is characterized by constitutive constraints and conservation laws. Nonlocal phenomena emerge from local interactions leading to new functionality via cooperation.

Remarkably, the framework uses a mathematical language that is abstract enough to include the gauge theories of fundamental interactions, the dynamics of space-time, basic life processes, and—going beyond the material properties of the world—even propositional logic. With this brief outline of Mack's proposal, we refer the interested reader to the original reference and references therein.

References

Anderson, P.W.: More and Different, p. 359. World Scientific, Singapore (2011)

Bartelmann, M.: Structure formation in the universe. In: Meyer-Ortmanns, H., Thurner, S. (eds.) Principles of Evolution: From the Planck Epoch to Complex Multicellular Life, pp. 189–204. Springer, Berlin (2011)

Bhattacharya, N., et al.: Implementation of quantum search algorithm using classical fourier optics. Phys. Rev. Lett. **88**, 137901-1–137901-4 (2002)

---

[8] For an example in which the field theoretic formalism was used in connection with ecology, see again (Mobilia et al. 2007).

Bustamente, C., et al.: Ten years of tension: single-molecule DNA mechanics. Nature **421**, 423–427 (2003)

Ellwanger, U., et al.: Evolution equations for the quark-meson transition. Nucl. Phys. B **423**, 137–170 (1994)

Fratzl, P., Weinkammer, R.: Nature's hierarchical materials. Prog. Mater Sci. **52**(8), 1263–1334 (2007)

Freddolino, P.L., et al.: Molecular dynamics simulations of the complete satellite tobacco mosaic virus. Structure **14**, 437–449 (2006)

Frey, E.: Evolutionary game theory: theoretical concepts and applications to microbial communities. Phys. A **389**, 4265–4298 (2010)

Gierer, A., Meinhardt, H.: A theory of biological pattern formation. Kybernetik **12**, 30–39 (1972)

Glashow, S.L.: Partial symmetries of weak interactions. Nucl. Phys. **22**(4), 579588 (1961)

Gross, P., et al.: Quantifying how DNA stretches, melts and changes twist under tension. Nat. Phys. **7**, 731–736 (2011)

Grover, L.K.: Quantum mechanics helps in searching for a needle in a haystack. Phys. Rev. Lett. **78**, 325–328 (1997)

Hehl, F.W., et al.: General relativity with spin and torsion: foundations and prospects. Rev. Mod. Phys. **48**, 393–416 (1976)

Kadanoff, L.P.: The application of renormalization group techniques to quarks and strings. Rev. Mod. Phys. **49**, 267–296 (1977)

Khaleel, M.A.: Scientific grand challenges: forefront questions in nuclear science and the role of high performance computing. Report from the workshop, 26–28 Jan 2009, report number: PNNL-18739, pp. 1–255 (2009). doi:10.2172/968204

Kibble, T.W.B.: Lorentz invariance and the gravitational field. J. Math. Phys. **2**, 212–222 (1961)

Kogut, J., Wilson, K.: The renormalization group and the ϵ-expansion. Phys. Rep. **12C**, 75–199 (1974)

Mack, G.: Physical principles, geometrical aspects and locality of gauge field theories. Fortschritte der Physik **81**, 135–185 (1981)

Mack, G.: Universal dynamics, a unified theory of complex systems, emergence, life and death. Commun. Math. Phys. **219**, 141–178 (2001)

May, R.M., Leonard, W.J.: Nonlinear aspects of competition between three species. SIAM J. Appl. Math. **29**, 243–253 (1975)

Meyer-Ortmanns, H., Reisz, T.: Principles of phase transitions in particle physics. World Scientific Lecture Notes in Physics, vol. 77. World Scientific, Singapore (2007)

Mobilia, M., Georgiev, IT., U.C.T.: Phase transitions and spatio-temporal fluctuations in stochastic lattice Lotka-Volterra models. J. Stat. Phys. **128**(1), 447–483 (2007)

Murray J.D.: A pattern formation mechanism and its application to mammalian coat markings. In: 'Vito Volterra' Symposium on Mathematical Models in Biology. Academia dei Lincei, Rome, Dec 1979. Lecture Notes in Biomathematics, vol. 39, pp. 360–399. Springer, Berlin (1980)

Murray, J.D.: A pre-pattern formation mechanism for animal coat markings. J. Theor. Biol. **88**, 161–199 (1981)

Murray, J.D.: Mathematical Biology II: Spatial Models and Biomedical Applications. Springer, Berlin (1993)

O'Raifeartaigh, L.: The Dawning of Gauge Theories. Princeton University Press, Princeton (1997)

Patel, A.: Quantum algorithms and the genetic code. Pramana **56**, 365–376 (2001)

Ramond, P.: Field Theory: A Modern Primer. Addison-Wesley, Redwood City (1989)

Rindler, W.: Essential Relativity. Van Nostrand Reinhold Company, New York (1969)

Salam, A.: Weak and electromagnetic interactions. In: Svartholm, N. (ed.) Elementary Particle Physics: Relativistic Groups and Analyticity, vol. 8, pp. 367–377. Proceedings of the nobel symposium (1968)

Sciama, D.W.: Recent Developments in General Relativity. Pergamon Press, Oxford (1962)

Turing, A.M.: The chemical basis of morphogenesis. Philos. Trans. R. Soc. Lond. **B327**, 32–72 (1952)

Wang, R., Gupta, H.S.: Deformation and fracture mechanisms of bone and nacre. Ann. Rev. Mat. Res. **41**, 41–73 (2011)

Weinberg, S.: A model of leptons. Phys. Rev. Lett. **19**, 1264–1266 (1967)

Weyl, H.: Space, Time, Matter. Springer, Berlin (1922)

Wilson, K.G.: The renormalization group: critical phenomena and the kondo problem. Rev. Mod. Phys. **47**, 773–840 (1975)

Yakusheich, L.V.: Nonlinear Physics of DNA. Wiley, Chichester (1998)

Yang, C.N., Mills, R.L.: Conservation of isotopic spin and isotopic gauge invariance. Phys. Rev. **96**, 191–195 (1954)

Yeomans, J.M.: Statistical Mechanics of Phase Transitions. Clarendon Press, Oxford (1992)

Chapter 3
On the Relation Between the Second Law of Thermodynamics and Classical and Quantum Mechanics

Barbara Drossel

3.1 Introduction

This article is devoted to the relation between the second law of thermodynamics, which applies to closed macroscopic systems consisting of an extremely large number of particles, such as liquids or gases, and classical or quantum mechanics, which are theories that describe systems of interacting particles on a microscopic level. The second law of thermodynamics states that in a closed system entropy increases until it reaches a maximum. Then the system has reached equilibrium. Entropy is a property of a "macrostate", which is characterized by measurable variables such as density $\rho(\vec{x})$ or magnetization $M(\vec{x})$, these being sums or averages over many particles. The entropy is assumed to be proportional to the logarithm of the number of different "microstates" that correspond to such a macrostate. Based on the concept of microstates, the second law of thermodynamics (and all other relations of thermodynamics) can be obtained from statistical mechanics with its basic axiom that, in a closed system in equilibrium, all microstates occur with equal probability. This means that transition probabilities between microstates are such that in the long run no state is preferred.

Now, the microscopic description of a many-particle system in terms of classical or quantum mechanics differs in two fundamental ways from the statistical mechanics description which entails the second law of thermodynamics. First, classical mechanics and quantum mechanics are deterministic theories. Given the initial state of a system, these theories determine its future time evolution. In contrast, statistical mechanics is a stochastic theory, with probability being an important concept. Second, classical mechanics and quantum mechanics are time reversible. If a given trajectory is a solution of Newton's laws, the time inverted

B. Drossel (✉)
Institut für Festkörperphysik, Technische Universität Darmstadt,
Hochschulstr. 6, 64289 Darmstadt, Germany
e-mail: drossel@fkp.tu-darmstadt.de

© Springer-Verlag Berlin Heidelberg 2015
B. Falkenburg and M. Morrison (eds.), *Why More Is Different*,
The Frontiers Collection, DOI 10.1007/978-3-662-43911-1_3

trajectory is also a solution, because Newton's laws do not change under time reversal, involving as they do the second derivative with respect to time. Similarly, if a wave function $\psi(\vec{x}, t)$ is a solution of the Schrödinger equation, its complex conjugate $\psi^*(\vec{x}, t)$ is a solution of the time reversed Schrödinger equation, giving exactly the same physical properties, since observables depend only on the absolute value of the wave function. In contrast, the second law of thermodynamics makes a fundamental distinction between the two directions of time. The entropy increase occurs only in the forward time direction.

Because of these two fundamental differences, the question arises if and how the second law of thermodynamics (and statistical mechanics in general) can be derived from classical or quantum mechanics. Many textbook authors assume that in principle a macroscopic system is on the microscopic level fully and correctly described by deterministic, time-reversible laws. Of course, if such a microscopic description is complete, it must somehow contain all the properties that are perceived in thermodynamic systems that consist of the order of 10^{23} particles. Consequently, the irreversible character of the second law is ascribed by these textbook authors to our inability to obtain knowledge of the precise microscopic state of the system, combined with special initial conditions for the macroscopic, observable quantities of the system. This is the so-called ignorance interpretation of probability. Our inability to know the microstate of the system and hence to predict its future evolution is aggravated by the fact that no system is fully isolated from the rest of the world. This means that the future evolution of a system is influenced by its environment. In order to see the deterministic character of a system's time evolution, one would need to know the state of its environment, which in turn depends on still another environment, etc. Since it is impossible to include in our calculations such a wider environment, which might consist of the entire universe, some textbook authors argue that we have no choice but to describe our system using the concept of probabilities.

At this point it becomes clear that the belief that the time evolution of a many-particle system is deterministic on the microscopic level is a metaphysical belief. It cannot, even in principle be shown to be correct. It is a starting assumption on which the subsequent considerations are based, and it is not the result of scientific observations. In order to assess how reasonable this basic assumption is, one has to explore its logical consequences. In fact, the debate on how to relate the probabilities of statistical mechanics to a microscopic view is as old as the theory itself, starting from the Boltzmann–Zermelo debate and continuing until today (see for instance the contribution by Jos Uffink in Beisbart and Hartmann 2011).

In the following, I will argue that the second law of thermodynamics cannot be derived from deterministic time-reversible theories such as classical or quantum mechanics. This means that even simple macroscopic equilibrium systems such as gases, crystals, or liquids, cannot be fully explained in terms of their constituent particles, or, to use the wording in the title of this book, it means that "more is different". In the next section, I will challenge the basic assumption of many textbook authors that classical or quantum mechanics can provide an accurate, comprehensive microscopic description of a thermodynamic system. Then I will show that all so-called "derivations" of the basic concepts of statistical mechanics and in particular of the second law of thermodynamics make assumptions that go beyond classical or quantum mechanics, often tacitly.

3.2 The Mistaken Idea of Infinite Precision

In classical mechanics, the state of a system can be represented by a point in phase space. The phase space of a system of N particles has $6N$ dimensions, which represent the positions and momenta of all particles. Starting from an initial state, Newton's laws, in the form of Hamilton's equations, prescribe the future evolution of the system. If the state of the system is represented by a point in phase space, its time evolution is represented by a trajectory in phase space. However, this idea of a deterministic time evolution represented by a trajectory in phase space can only be upheld within the framework of classical mechanics if a point in phase space has infinite precision. If the state of a system had only a finite precision, its future time evolution would no longer be fixed by the initial state, combined with Hamilton's equations. Instead, many different future time evolutions would be compatible with the initial state. In practice, it is impossible to know, prepare, or measure the state of a system with infinite precision. This would require a brain or another computing device that can store an infinite number of bits, and such a thing does not exist in a finite universe. Here, we see once again that the belief that classical mechanics can provide a valid microscopic description of a thermodynamic system is a metaphysical belief.

In quantum mechanics, the state of a system is represented by a wave function, and its time evolution by the Schrödinger equation. Now, in order for the Schrödinger equation to fully predict the future evolution of a quantum mechanical system, it is not just a point that must be specified with infinite precision, but a complex-valued function of $3N$ variables, because the wave function is complex and depends on the positions of all particles. Even the "simple" enterprise of calculating the ground state wave function of a many-electron system fails completely for more than 1,000 particles, says Walter Kohn, the father of density functional theory, in his Nobel lecture, quoting his teacher Van Vleck: "The many-electron wave function is not a legitimate scientific concept for $N > 1000...$ because the wave function can neither be calculated nor recorded with sufficient accuracy" (Kohn 1999). This illustrates once again that the idea that the state of a system has infinite precision is a metaphysical assumption.

History and philosophy of science have taught us that the idealizations that are contained in the theories of physics should not be taken as a faithful and perfect reflection of reality. In classical mechanics, these idealizations include certain concepts of space and time, in addition to a deterministic worldview. The belief that classical mechanics is a faithful and perfect reflection of physical reality was shattered 100 years ago, when the theory of relativity showed it to be an approximation that is pretty good when velocities are far below the velocity of light and when gravitational fields are weak enough to mean that the curvature of space cannot be perceived. Furthermore, the advent of quantum mechanics made it clear that classical mechanics is not valid on atomic length scales. In particular, the uncertainty principle states that a point in phase space cannot have infinite precision. This has drastic consequences because the discovery of chaos demonstrated that in many systems a limited precision of the initial state leads to complete uncertainty in its future time evolution beyond a short time horizon.

In quantum mechanics, we know that the Schrödinger equation is a good description of a system only when relativistic effects and radiation effects can be neglected. In addition to these reasons for considering wave functions and the Schrödinger equation only as an approximate description of reality, quantum mechanics poses a far more fundamental problem. It has itself a part that is inherently stochastic and irreversible, namely quantum measurement. With respect to the outcome of a measurement, only probabilities can be given. Furthermore, a measurement process, which includes irreversible changes in the macroscopic measurement device, does not occur backwards in time, thus making a distinction between past and future. Landau and Lifschitz suggest in their textbook on statistical physics that the irreversibility of the measurement process is related to the irreversibility of the second law of thermodynamics. Despite claims made to the contrary by some scientists, the measurement process has not been satisfactorily explained in terms of the deterministic evolution of a many-particle wave function according to the Schrödinger equation (Schlosshauer 2004).

All these well known limitations of classical mechanics and quantum mechanics support my argument that the idea of infinite precision, which follows from these theories if they are an exact reflection of reality, is a very questionable concept. The belief that the time evolution of a thermodynamic system is deterministic and reversible on the microscopic level is a metaphysical assumption that presents a variety of problems. As soon as the metaphysical assumption of infinite precision is abandoned, the present state of a system, combined with the microscopic laws of classical or quantum mechanics, does not fully specify the future time evolution. Therefore, additional laws are needed to specify which of the possible types of time evolution are adopted by the system.

I therefore join those textbook authors that maintain that statistical mechanics is a field of physics in its own right, which is not contained in deterministic, time-reversible microscopic theories. Statistical mechanics has its own axiom, namely the axiom of "equal a priori probabilities". This axiom states that in a closed system in equilibrium all microstates occur with equal probability. From this axiom, all of statistical mechanics can be derived.

In the next section, we will show that even those scientists who "derive" the law of equal a priori probabilities and the second law of thermodynamics from classical mechanics nevertheless make assumptions that go beyond classical mechanics. Then, we will look briefly at other approaches to the relation between classical and statistical mechanics, which openly employ additional assumptions and thus admit that statistical mechanics cannot be shown to be contained in classical mechanics.

In Sect. 3.3, we make a short survey of the different possible ways to relate statistical mechanics to quantum mechanics, encountering similar and even worse
challenges than for classical mechanics.

3.3 From Classical Mechanics to Statistical Mechanics

3.3.1 The Standard Argument

As already mentioned, every "derivation" of statistical mechanics from classical
mechanics starts from the metaphysical assumption that classical mechanics can
provide a complete and deterministic description of a 10^{23}-particle system. An
important part of this assumption is the idea that the state of a system has infinite
precision. We will now start from this assumption and develop the usual arguments
that can be found in many textbooks. We will see that in fact the deterministic
assumption is not followed through to the end, but that probabilistic assumptions,
which are foreign to classical mechanics, creep in, often without being perceived as
such.

An important concept when relating classical mechanics to statistical mechanics
is "quasi-ergodicity": The concept of quasi-ergodicity means that a "typical" trajectory in phase space comes arbitrarily close to every point on the energy shell if
one waits long enough. "Typical" are all trajectories with the exception of a few
special trajectories, the initial points of which have measure zero in the energy shell.
Unstable periodic orbits belong to this special class of trajectories. The "energy
shell" is the subspace of phase space that has the energy of the initial state. Due to
the Hamiltonian character of time evolution, energy is conserved. Due to the
strongly chaotic character, there are no other conserved quantities and supposedly
no islands of regular dynamics. If we lay a grid with some cell size ε^{6N-1} over the
energy shell, a typical trajectory will have visited all cells of the grid after a time
that depends on ε. Since there is no reason to prefer any part of the energy shell, all
parts will be visited equally often on average. This is one way of stating that all
microstates of the system occur with equal probability.

It is also instructive to consider the time evolution of an ensemble of initial
states, all of which lie in a small compact volume in phase space. With time, the
volume becomes deformed according to Liouville's equation, but its size does not
change. Due to the stretching and folding process of chaotic motion, finer and finer
filaments of the volume will penetrate to more and more cells of our phase space
grid. After some time, all cells will contain a small part of the original volume in the
form of very fine filaments. If the size of the initial droplet represents the precision
of our knowledge of the initial state, we cannot say at all in which cell the system
will be after this time. It can be anywhere with equal probability. This leads again to
the basic axiom of statistical mechanics.

In order to justify the second law of thermodynamics, an additional consideration is needed: most cells in the energy shell correspond to the macrostate with the highest entropy. This is a consequence of the huge dimension of phase space. Therefore, the argument goes, even if one starts from a low-entropy initial state, after a short time the system will reach "typical" cells in phase space, which have maximum entropy.

3.3.2 The Problems with the Standard Argument

There are two important problems with the argument outlined in the previous subsection, both of which are closely tied to the underlying idea that the state of a system corresponds to a point in phase space.

First, as pointed out by several textbook authors, the concept of quasi-ergodicity poses problems, because the time required to visit every cell in phase space is considerably longer than the lifetime of the universe. During the short time of an experiment, where we observe a system to reach equilibrium, only a vanishingly small fraction of phase space can be visited by the system. Therefore, the fundamental theorem of statistical mechanics, i.e., the theorem of equal a priori probabilities, cannot be derived from classical mechanics in this way. Fortunately, quasi-ergodicity is not required to justify the second law of thermodynamics, since we only need to argue that each cell in the energy shell is not far from maximum entropy cells, which are by far the most numerous types of cells.

However, when claiming that a trajectory will after some time "most likely" or "virtually certainly" be in cells with maximum entropy, one makes a probabilistic argument, which cannot be justified on the basis of classical mechanics alone, and this is the second problem with the standard derivation sketched in the previous subsection. Stated somewhat differently and in more detail, the probabilistic argument must be understood as follows. Given an initial state of the system, consider all future time evolutions that are compatible with this initial state within the given precision. Assuming that all these time evolutions happen with the same probability, and given the fact that the vast majority of these time evolutions show an increase in entropy towards equilibrium, the system will show an increase in entropy and approach equilibrium. Without the possibility of different time evolutions, probabilistic statements make no sense. A strictly deterministic world with infinite-precision phase space points leaves no freedom to "choose" the most probable time evolution, because the initial state fully contains the future time evolution.

A time evolution that does not agree with the "most probable" behavior is also compatible with classical mechanics, as one can conclude from the fact that, by going backwards in time, a system will arrive at the initial state at which it started. Initial states with an entropy that is smaller than that of equilibrium occur for instance when milk is poured into coffee or when the dividing barrier between two different gases is removed. By going backwards in time, starting from equilibrium, the entropy of these systems would decrease, and the time evolution would tend to a highly "improbable" state. We conclude that a time evolution towards a state of lower entropy (now going again forward in time) would in no way contradict classical mechanics. For such a case, one would have to conclude from a deterministic point of view that the initial state, even though we cannot resolve it to the required high precision, was one of the "special" initial states that tend to low-entropy states at some later time. Such states lie dense in the energy shell even though they have measure zero.

3.3.3 An Alternative View

By abandoning the idea that a position in phase space has infinite precision, all the problems raised in the previous subsection are solved. In order to show this, we start now from the assumption that a point in phase space has limited precision, and explore its consequences. Because Hamilton's equations do not unequivocally fix the future time evolution, we have to combine the idea of finite precision of phase space points with a rule that specifies which type of time evolution out of the possible ones is taken by the system. This rule is the rule of equal probabilities for (or at least of a smooth probability distribution over) those evolutions that are compatible with the initial state.

The assumption that points in phase space have limited precision is logically very satisfying for several reasons. First, as already mentioned, it creates room for employing probabilistic rules, which are the basis of statistical mechanics. Second, a finite precision of phase space points is all that is ever needed for "deriving" statistical mechanics from classical mechanics. In particular, when discussing the concept of quasi-ergodicity one always resorts to considering phase space with a certain resolution, given by the mesh size of the grid mentioned above. Third, a finite precision of phase space permits a system to reach equilibrium within a short time and to truly forget the past. With infinite precision, the time required to visit all cells of the phase space grid is proportional to the number of cells, and many orders of magnitude longer than the age of the universe. When the initial state and all future states have only finite precision, which we can take to be identical to the cell size, the number of cells which the system can reach within a certain time is greater than 1. This means that the number of cells that can be reached from the initial state increases exponentially in time, leading to the conclusion that, after a short time, the system could be anywhere on the energy shell. Furthermore, a true equilibrium state should carry no trace of the past. However, with infinite precision, the state of a system would always be uniquely related to a predecessor state at a previous moment in time. With finite precision, the number of possibly predecessor cells increases exponentially with the length of time over which one looks back. When this time interval becomes long enough, the system could have been anywhere on the energy shell, and the initial state is completely forgotten.

All these considerations can be made using classical mechanics alone in the attempt to reconcile it with the second law of thermodynamics. Of course, it is very satisfying to know that quantum mechanics confirms the suggestion that points in phase space have only limited precision. The uncertainty relation makes it impossible to fix simultaneously the momenta and positions of particles to infinite precision. Furthermore, quantum mechanics fixes the "mesh size" for the phase space grid. A cell in phase space has size $\hbar^{3N}$.

An important conclusion from these considerations is that the time evolution of a thermodynamic system is underdetermined by classical mechanics if the idea of infinite precision is abandoned. An additional law is required that specifies which type of time evolution is taken. This law is the law of equal probabilities, leading naturally to the second law of thermodynamics. Thus, the second law of thermodynamics is an emergent law in the strong sense; it is not contained in the microscopic laws of classical mechanics.

3.3.4 Other Routes from Classical Mechanics to the Second Law of Thermodynamics

There exist other approaches to statistical mechanics, which expressly make additional assumptions that are not part of classical mechanics. In their textbook on statistical physics, Landau and Lifschitz reject the idea of quasi-ergodicity of a large system because of the impossibility that a system visit even a small part of phase space within the duration of an experiment. Instead, they divide the system into many small subsystems and note that observables are sum variables over all these subsystems. By assuming that these subsystems are statistically independent, equal a priori probabilities and the second law of thermodynamics can be obtained. By assuming statistical independence of the subsystems they make an assumption that is similar in spirit to the idea of the previous subsections. They assume that all future time evolutions compatible with the initial state are equally probable: statistical independence of subsystems means that the influence of one subsystem on a neighboring one does not depend on the specific microscopic state of the subsystem. Rather, the influence on the neighbor is a "typical" influence, as if the neighbor were in a random state. Specific correlations between subsystems or processes that have taken place in the past are irrelevant for equilibrium behavior.

A similar type of assumption underlies Boltzmann's equation. This equation describes the time evolution of a gas of particles which goes towards an equilibrium state. Since this equation appears very plausible, one is tempted to forget that it contains assumptions that are not part of classical mechanics. In particular, this equation relies on the assumption that correlations due to past processes are irrelevant. Bolzmann's equation can be derived from Hamilton's equations by making a few simplifications. It is based on the density of particles $f(\vec{p}, \vec{q}, t)$ in 6-dimensional phase space. This phase space is the phase space of one particle, and the state of the system is represented by N points in this phase space. Time evolution is determined by collisions between particles and by free motion in-between. If no external potential is included, Boltzmann's equation reads

$$\frac{\partial f(\vec{p},\vec{q},t)}{\partial t} + \dot{\vec{q}} \cdot \frac{\partial f(\vec{p},\vec{q},t)}{\partial \vec{q}} = \int d^3p_2 \int d^3p_3 \int d^3p_4$$
$$W(\vec{p},\vec{p}_2;\vec{p}_3,\vec{p}_4)[f(\vec{p}_3,\vec{q},t)f(\vec{p}_4,\vec{q},t) - f(\vec{p},\vec{q},t)f(\vec{p}_2,\vec{q},t)].$$
$$(3.1)$$

Collisions between particles with momenta $\vec{p}_3$ and $\vec{p}_4$ lead to the momenta $\vec{p}$ and $\vec{p}_2$, or vice versa. The function W contains the cross-section for such collisions. The collision term depends only on the products of one-particle densities f, which means that the probabilities for particles being at the position $\vec{q}$ are assumed to be independent from each other. Correlations between particles, which are created by collisions, are thus neglected. The success of Boltzmann's equation justifies this assumption, and it means that a detailed memory of past processes is not required for correctly predicting the future time evolution. Once again, we find that the future time evolution is assumed to be a "typical" time evolution, which results when the present microstate is a random state compatible with the observables, i.e., with the function $f(\vec{p},\vec{q})$. It is well known that Boltzmann's equation leads to a decrease in time of the so-called H-function, which is the integral of $(f \log f)$ over phase space. The system approaches equilibrium, where the H-function has its minimum, which is equivalent to the entropy having its maximum.

3.4 From Quantum Mechanics to Statistical Mechanics

There are essentially four approaches to connecting quantum mechanics with statistical mechanics. The first considers an N-particle wave function of a closed system, the second includes the interaction with the environment via a potential, the third models the environment as consisting of many degrees of freedom, and the fourth treats quantum mechanics as an ensemble theory and views statistical mechanics as being part of quantum mechanics. Here we briefly discuss all four approaches, their achievements, and their shortcomings. We shall again see that in the first three approaches additional assumptions must be made about correlations being absent and time evolutions showing "typical" behavior.

3.4.1 The Eigenstate Thermalization Hypothesis

A system of N interacting particles in a potential well has chaotic dynamics in classical mechanics. In quantum mechanics, its eigenstate wave functions look very random. When dividing the volume of the potential well into small subvolumes, we can expect an eigenfunction of the Hamiltonian to have equal particle density in all subvolumes. The eigenstate thermalization hypothesis suggests that in fact all expectation values of observables, evaluated with an eigenfunction, correspond to the thermal average in equilibrium (Deutsch 1991; Srednicki 1994; Rigol et al. 2008). Using an appropriate superposition of eigenstates, one can generate an initial state that is far from equilibrium, and assuming that the expansion coefficients are such that they do not lead to special superpositions at later times, the system is bound to evolve towards an "equilibrium" state, where the observables have the corresponding values.

The problems with this approach are twofold. First, in order to obtain the "typical" time evolution towards equilibrium, we must make the assumption that there are no correlations in the expansion coefficients that might lead to special, low-entropy states at later times. This is the same type of assumption that is made in the approaches to classical mechanics that we discussed in the previous section. Second, this approach does not give the kind of density matrix that represents a thermodynamic equilibrium. The density matrix of this system is that of a pure state, with diagonal elements that are constant in time and that correspond to the weights of the different eigenfunctions in the initial state. In contrast, the density matrix at equilibrium is that of a mixed state, which can be represented as a diagonal matrix in the basis of energy eigenstates, with all entries being identical.

In the previous section, we argued that the problems that arise when reconciling classical mechanics with the second law of thermodynamics are resolved when we abandon the idea of infinite precision. It appears to me that the quantum mechanical time evolution of a closed system can be reconciled with statistical mechanics in a similar way, by abandoning the idea of infinite precision of a wave function. Only if a wave function has finite precision does it make sense to say that the expansion coefficients have the "most likely" property of leading to no special microstates at later times. Furthermore, if a wave function has finite precision, it can be compatible with many pure and mixed (infinite-precision) states, i.e., with many different density matrices, and with many possible future time evolutions. Which time evolution is actually followed by the system has then to be fixed by additional laws.

3.4.2 Interaction with the Environment Through a Potential

This approach is due to Felix Bloch (1989), and it starts by taking the Hamilton operator to be that of an isolated system, H_0, plus an interaction potential that represents the effect of the environment. Since no system can be completely isolated from the rest of the world, there is always some external influence. Starting from an initial state $|\Psi, 0\rangle = \sum_n c_n |n\rangle$, the wave function becomes at later times

$$|\Psi, t\rangle = \sum_n c_n(t) e^{-iE_n t/\hbar} |n\rangle. \tag{3.2}$$

Assuming that the interaction with the environment does not change or barely changes the energy of the system, the energy eigenvalues E_n all lie within a very small interval. The time evolution of the coefficients c_n can be calculated to be

$$i\hbar \frac{\partial c_m}{\partial t} = \sum_n V_{mn}(t) c_n(t), \tag{2.3}$$

with

$$V_{nm}(t) = \langle m|V|n \rangle \, e^{-i(E_n - E_m)t/\hbar}. \tag{2.4}$$

Because the vector c_n is normalized to 1 ($\sum_n |c_n|^2 = 1$), its tip moves on the unit sphere. If the system is ergodic in a suitable sense, the tip of the vector will come arbitrarily close to every point on this sphere. There is a very close analogy between the trajectory $c_n(t)$ and the trajectory of a classical mechanical trajectory in phase space. All the reasoning made above in the context of classical trajectories therefore applies also to $c_n(t)$. By abandoning the idea of infinite precision of a quantum mechanical state, the problem that ergodicity requires incredibly long time periods would be resolved, along with the problem that a completely deterministic time evolution leaves no room for appeals to the "most likely" behavior.

3.4.3 Coupling to an Environment with Many Degrees of Freedom

Since the 1980s, a series of very fruitful investigations have been performed on the time evolution of a quantum mechanical system that interacts with an environment comprising many degrees of freedom, all of which are also modelled quantum mechanically, for instance as harmonic oscillators. The time evolution of the system and all environmental degrees of freedom is taken to be a unitary time evolution according to the Schrödinger equation. Accordingly, the density matrix of the full system, which includes the environmental degrees of freedom, is that of a pure state. However, when focussing on the system of interest, the trace over the environmental degrees of freedom is taken, leading to a reduced density matrix, which is generally that of a mixed state.

One important application of this procedure is quantum diffusion (Caldeira and Leggett 1983), where a particle is coupled to the environmental degrees of freedom via its position. Using the path integral formalism and taking in the end the limit $\hbar \to 0$, one finds that the probability density that the particle has moved through a distance $\vec{r}$ during time t is described by the Fokker–Planck equation. Thus, the transition has been made from a quantum mechanical description to a classical, stochastic description. This type of phenomenon is called "decoherence".

Another important application of decoherence theory is that of quantum measurement. In this case, the system couples to the many degrees of freedom of the (macroscopic) measurement apparatus via the observable that is being measured, for instance the spin. In turns out that the reduced density matrix of the system becomes diagonal after a very short time (Zurek 1991; Zeh 2002). This means that it describes a classical probabilistic superposition of the different measurement outcomes. In a similar vein, one can argue that the density matrix of a thermodynamic system (or a small subsystem of it) will evolve to the density matrix of a classical superposition when the system is coupled to an environment that consists of many degrees of freedom, and when the trace is taken over these degrees of freedom.

A close look at these calculations reveals that they rely on two types of assumptions that are similar in spirit to the assumptions that we have discussed before. First, one must assume some kind of statistical independence or lack of special correlations between the variables that describe the environmental degrees of freedom. The reduced density matrix of the system becomes diagonal only if the non-diagonal elements, which contain products of a function of the amplitudes and phases of the different environmental degrees of freedom, decrease to zero. Second, it must be assumed that for all practical purposes the entanglement of the system with the environment, which is pushed out of sight by taking the trace over the environmental degrees of freedom, can be ignored. Such an entanglement would contain a full memory of the process that has taken place since the system started to interact with the environment. Now, we can argue once more that a finite precision of the wave function would solve these conceptual problems: the entanglement with the environment and the perfect memory of the past could vanish with time, and the lack of special correlations could be phrased in terms of the most probable or typical time evolution.

Critics of decoherence theory focus on its incomplete potential to explain the measurement process. If many experiments of the same type are performed, the diagonal entries of the density matrix tell correctly which proportion of experiments will show which measurement result. However, quantum mechanics is taken to be a theory that describes individual systems, not just ensembles of systems. This is not the topic of this article, but it leads us to the fourth approach to the relation between quantum and statistical mechanics.

3.4.4 Quantum Mechanics as a Statistical Theory that Includes Statistical Mechanics

A view of quantum mechanics that circumvents the need to reconcile the deterministic, time-reversible evolution of the Schrödinger equation with the stochastic, irreversible features of statistical mechanics, is the statistical interpretation (Ballentine 1970). In this interpretation, quantum mechanics is viewed as an ensemble theory, which gives probabilities for measurement outcomes. By extending this theory to include mixed states, statistical mechanics becomes part of quantum mechanics.

The price for this elegant solution of our problem is the incompleteness of the theory, because individual systems cannot be described by it. But once again, this is not the topic of this article.

3.5 Conclusions

By taking a close look at the different "derivations" of statistical mechanics, in particular of the second law of thermodynamics, from classical or quantum mechanics, we have seen that all these derivations make similar assumptions that go beyond the deterministic, time-reversible microscopic theory from which they start. All derivations assume that the time evolution of the system is "typical" in some sense. This means that the most likely type of time evolution of observable variables, given our knowledge of the system, does occur. If the time evolution of the system was deterministic on the microscopic level, our probabilistic statements about the system would merely be due to our ignorance of the precise microscopic state. However, in this case there would be no conclusive reason why the time evolution of the system should comply with our ignorance and take the "most likely" route, because other routes would also be compatible with our knowledge of the initial state. Furthermore, when looking into the past, the evolution backwards in time does not take the most likely route. We must therefore understand the rule of equal probabilities as an additional law that is required to describe the behavior of the system correctly. However, there would be no room for an additional law if the laws of classical or quantum mechanics did fully determine the time evolution of the system. We are therefore led to conclude that the description in terms of classical or quantum mechanics is only an approximate description, and that points in phase space or the wave function have only a limited precision. This line of reasoning is consistent with other philosophical and scientific arguments. From the philosophical point of view, the concept of infinite precision of a state is a metaphysical idea that cannot, even in principle, be tested. From a scientific point of view, we know that Newton's (or Hamilton's) equations of motion and the Schrödinger equation are only an approximation to reality. The fact that these two theories work so well for many applications can make us blind to the the possibility that their limited precision may have significant effects in systems that consist of macroscopic numbers of nonlinearly interacting particles. Such complex systems are therefore not simply the sum of their parts, but are governed by new laws that are not contained in a microscopic description.

References

Ballentine, L.E.: The statistical interpretation of quantum mechanics. Rev. Mod. Phys. **42**, 358–381 (1970)

Beisbart, C., Hartmann, S. (eds.): Probabilities in Physics, Oxford University Press, Oxford, (2011)

Bloch, F.: Fundamentals of Statistical Mechanics. Manuscript and Notes of Felix Bloch. In: Walecka, J.D. (eds.) Stanford University Press, Stanford (1989) (World Scientific 2000)

Caldeira, A.O., Leggett, A.J.: Path integral approach to quantum Brownian motion. Phys. A **121**, 587–616 (1983)

Deutsch, J.M.: Quantum statistical mechanics in a closed system. Phys. Rev. A **43**, 2046–2049 (1991)

Kohn, W.: Electronic structure of matter-wave functions and density functionals. Rev. Mod. Phys. **71**, 1253–1266 (1999)

Rigol, M., Dunjko, V., Olshanii, M.: Thermalization and its mechanism for generic isolated quantum systems. Nature **452**, 854–858 (2008)

Schlosshauer, M.: Decoherence, the measurement problem, and interpretations of quantum mechanics. Rev. Mod. Phys. **76**, 1268–1305 (2004)

Srednicki, M.: Chaos and quantum thermalization. Phys. Rev. E **50**, 888–901 (1994)

Zeh, H.D.: The wave function: It or bit. *arXiv:quant-ph/0204088v2* (2002)

Zurek, W.H.: Decoherence and the transition from quantum to classical. Phys. Today **44**, 36–44 (1991)

Chapter 4
Dissipation in Quantum Mechanical Systems: Where Is the System and Where Is the Reservoir?

Joachim Ankerhold

4.1 Introduction

Brownian motion, that is the fate of a heavy particle immersed in a fluid of lighter particles, is *the* prototype of a dissipative system coupled to a thermal bath with infinitely many degrees of freedom. The work by Einstein in (1905) developed a mathematical language to describe the random motion of the particle and uncovered the fundamental relation between friction, diffusion, and the temperature T of the bath. Half a century later, this seed had grown into the theory of irreversible thermodynamics (Landau and Lifshitz 1958), which governs the relaxation and fluctuations of classical systems near equilibrium. By that time, a new challenge had emerged, the quantum mechanical description of dissipative systems.

In contrast to classical Brownian motion, where right from the beginning the work by Einstein and Smoluchowski (Smoluchowski 1906) had provided a way to consider both weak and strong friction, the quantum mechanical theory could for a long time only handle the limit of weak dissipation. In this case the interaction between the "particle" and the "bath" can be treated perturbatively and one can derive a master equation for the reduced density matrix of the "particle" (Blum 1981). This approach has been very successful in quite a number of fields emerging in the 1950s and 1960s, such as nuclear magnetic resonance (Wangsness and Bloch 1953; Redfield 1957) and quantum optics (Gardiner and Zoller 2004). It turns out that, in this regime, conventional concepts developed in classical thermodynamics still apply, since the "particle" can be considered as a basically independent entity while the role of the reservoir is to induce decoherence and relaxation towards a thermal Boltzmann distribution solely determined (apart from temperature) by properties of the "particle".

J. Ankerhold (✉)
Institut für Theoretische Physik, Universität Ulm, Albert Einstein-Allee 11,
89069 Ulm, Germany
e-mail: joachim.ankerhold@uni-ulm.de

© Springer-Verlag Berlin Heidelberg 2015
B. Falkenburg and M. Morrison (eds.), *Why More Is Different*,
The Frontiers Collection, DOI 10.1007/978-3-662-43911-1_4

Roughly speaking, a dissipative quantum system can be characterized by three typical energy scales: an excitation energy $\hbar\omega_0$, where ω_0 is a characteristic frequency of the system, a coupling energy $\hbar\gamma$ to the bath, where γ is a typical damping rate, and the thermal energy $k_B T$. The weak coupling master equation is limited to the region $\hbar\gamma \ll \hbar\omega_0, k_B T$. This is the case whenever the typical linewidth caused by environmental interactions is small compared to the line separation and the thermal "Matsubara" frequency $2\pi k_B T/\hbar$.

It is thus to be expected that, for stronger damping and/or lower temperatures, quantum mechanical non-locality may have a profound impact on the system-("particle") bath correlation. With the further progress in describing the system-bath interaction non-perturbatively in terms of path integrals (Weiss 2008), it has turned out that this is indeed the case. In fact, quantum Brownian motion in this regime gives rise to collective processes that cannot be understood from those of the independent parts (emergence). Accordingly, the concept of "reduction" on which the formulation of open quantum systems is based, wherein one concentrates on a relevant system and keeps from the actual reservoir only its effective impact on this system, appears in a new light. In particular, while the procedure has been extremely successful, it forces us to abandon conventional perceptions about the role of what we consider as the "system" and what we consider as the "surroundings". The goal of the present contribution is to illustrate and discuss this situation.

To set the stage, I will start by briefly recalling the formulation of noisy classical dynamics and then proceed with a discussion about the general formulation of open quantum systems. The three specific examples to follow illustrate various facets of the intricate correlations between a system and its reservoir. As such they may also reveal new aspects of the concatenation between quantum system and observer.

4.2 Dissipation and Noise in Classical Systems

Energy dissipation in classical systems is a well-known phenomenon. A prominent example is the dynamics of a damped pendulum: starting initially from an elongated position, one observes an irreversible energy flow out of the system which finally brings it back to its equilibrium state. A more subtle situation is the diffusive motion of a small particle immersed in a liquid, also known as Brownian motion. As first pointed out by Einstein (1905), the stochastic movement of the small particle directly reflects scattering processes with the molecules of the thermal environment. In a stationary situation, the energy gained/lost in each scattering event is balanced by an energy flow into/out of this thermal reservoir. A brute force description of Brownian motion would start from Newton's equation of motion for the whole compound system, small particle and molecules in the liquid. From a purely practical point of view, this is completely out of reach, not only due to the enormous number of degrees of freedom but also owing to the fact that microscopic details of the molecule–molecule and molecule–particle interaction are not usually known in detail. However, even if this information were available and even if we were able to simulate the complex dynamics numerically, what would we learn if we were only interested in the motion of the immersed particle? The relevant information would have to be extracted from a huge pile of data and the relevant mechanism governing the interaction between particle and liquid would basically remain hidden.

Hence, the standard approach is based on Newton's equation of motion for the relevant particle augmented by forces describing both dissipative energy flow and stochastic scattering. For a one-dimensional particle of mass m and position q moving in a potential field $V(q)$, this leads in the simplest case to a so-called Langevin equation (Risken 1984):

$$m\ddot{q}(t) + V'(q) + m\gamma\dot{q}(t) = \xi(t), \tag{4.1}$$

where dots denote time derivatives and $V'(q) = dV/dq$. The impact of the reservoir only appears through the friction rate γ and the stochastic force $\xi(t)$, which has the properties

$$\langle\xi(t)\rangle = 0, \quad \langle\xi(t)\xi(t')\rangle = 2m\gamma k_B T\delta(t - t') \quad \text{(white noise)}.$$

This latter relation is known as a *fluctuation-dissipation theorem*, reflecting the fact that energy dissipation and stochastic scattering are inevitably connected since they have the same microscopic origin. A more realistic description takes into account the fact that the back-action of the reservoir on the system dynamics is time-retarded (colored noise), thus turning the constant friction rate into a time-dependent friction kernel $\gamma\dot{q}(t) \rightarrow \int_0^t ds\,\gamma(t - s)\dot{q}(s)$ (generalized Langevin equation). Anyway, the main message here is that, as long as we observe only the relevant particle, the impact of the reservoir is completely described by at least two macroscopic parameters, namely, the friction constant and the temperature, which can be determined experimentally. The complicated microscopic dynamics of the surrounding degrees of freedom need not be known.

4.3 Dissipative Quantum Systems

In contrast to the situation for classical systems, the inclusion of dissipation/fluctuations within quantum mechanics is much more complicated (Weiss 2008; Breuer and Petruccione 2002). A quantization of the classical Langevin Eq. (4.1) in terms of Heisenberg operators together with the quantum version of the fluctuation-dissipation theorem only applies for strictly linear dynamics (free particle, harmonic oscillator) and under the assumption that system and reservoir are initially independent. The crucial problem is that quantum mechanically the interaction between system and reservoir leads to a superposition of wave functions and thus to entanglement. This is easily seen when one assumes that the total compound system is described by a Hamiltonian of the form

$$H = H_S + H_I + H_R, \tag{4.2}$$

where a system H_S interacts via a coupling operator H_I with a thermal reservoir H_R. Whatever the structure of these operators, their pairwise commutators will certainly not all vanish. Typically, one has $[H_S, H_R] = 0$ while $[H_S, H_I] \neq [H_R, H_I] \neq 0$ for non-trivial dynamics to emerge. Accordingly, the operator for thermal equilibrium of the total structure, viz.,

$$W_\beta \sim \exp[-\beta(H_S + H_I + H_R)],$$

does not factorize (in contrast to the classical case). In a strict sense, the separation between the system and its environment no longer actually exists. Stating this in the context of the system-observer situation in quantum mechanics: the presence of an environment acts like an observer continuously probing the system dynamics.

The question is thus: how can we identify the system and reservoir from the full compound? The answer basically depends on the interests of the observer, and often simply on the devices available for preparation and measurement. Practically, one focuses on a set of observables $\{O_k\}$ associated with a specific sub-unit which in many cases coincides with the observables of a specific device that has been prepared or fabricated. The rest of the world remains unobserved. This then defines what is denoted as H_S in (4.2). Time dependent mean values follow from $\langle O_k \rangle = \mathrm{Tr}_S\{O_k \rho(t)\}$, where the reduced density operator

$$\rho(t) = \mathrm{Tr}_R\{U(t,0)W(0)U(t,0)^\dagger\} \tag{4.3}$$

is determined from the full time evolution $U(t,0) = \exp(-iHt/\hbar)$ of an initial state $W(0)$ of the full compound by averaging over the unobserved reservoir degrees of freedom. Conceptually, this is in close analogy to the classical Langevin Eq. (4.1) on the level of a density operator, with the notable difference though that a consistent quantization procedure necessitates knowledge of a full Hamiltonian (4.2). For the system part this may be obvious, but it is in general extremely challenging, if not impossible, for the reservoir and its interaction with the system.

Progress is made by recalling that what we defined as the surroundings typically contains a macroscopic number of degrees of freedom and, since it is not directly prepared, manipulated, or detected, basically stays in thermal equilibrium (Weiss 2008). Large heat baths, however, display Gaussian fluctuations according to the central limit theorem. A very powerful description applying to a broad class of situations then assumes that a thermal environment consists of a quasi-continuum of independent harmonic oscillators linearly coupled to the system, i.e.,

$$H_R + H_I = \sum_k \frac{p_k^2}{2m_k} + \frac{m_k \omega_k^2}{2} \left(x_k - \frac{c_k}{m_k \omega_k^2} Q \right)^2. \tag{4.4}$$

Here Q denotes the operator of the system through which it is coupled to the bath (pointer variable). For systems with a continuous degree of freedom, it is typically given by the position operator (or a generalized position operator). The system-bath interaction is written in a translational invariant form so that the reservoir only affects the system dynamically in a similar way to what happens in the classical case (4.1). In fact, the classical version of this model reproduces the Langevin equation, implying that the influence of the reservoir on the system is completely determined by the temperature T and the spectral distribution of the bath oscillators, viz.,

$$J(\omega) = \frac{\pi}{2} \sum_k \frac{c_k^2}{2m_k \omega_k} \delta(\omega - \omega_k).$$

A continuous distribution $J(\omega)$ ensures that energy flow from the system into the bath occurs irreversibly (Poincaré's recurrence time tends to infinity). Each oscillator may only weakly interact with the system, but the effective impact of the collection of oscillators may still capture strong interaction. For instance, in the case of so-called Ohmic friction corresponding classically to white noise, one has $J(\omega) = m\gamma\omega$.

The bath force $\xi = \sum_k c_k x_k$ acting on the system obeys Gaussian statistics and is thus completely determined via its first moment $\langle \xi(t) \rangle = 0$ and its second moment $K(t) = \langle \xi(t)\xi(0) \rangle$. As an equilibrium correlation, it obeys the quantum fluctuation-dissipation theorem, namely,

$$\tilde{K}(\omega) = 2\hbar J(\omega)\left(1 - e^{-\omega\hbar\beta}\right)^{-1}, \tag{4.5}$$

where $\tilde{K}(\omega)$ denotes the Fourier transform of $K(t)$ and $\beta = 1/k_B T$. One sees that for Ohmic dissipation $J(\omega) = m\gamma\omega$, and in the high temperature limit $\omega\hbar\beta \to 0$, the correlation becomes a constant $\tilde{K}(\omega) = 2m\gamma k_B T$ and thus describes the white noise known from classical dynamics (4.1). In the opposite limit of vanishing temperature $\omega\hbar\beta \to \infty$, however, one has $\tilde{K}(\omega) = 2m\gamma\hbar\omega$, which gives rise to an algebraic decay $K(t) \propto 1/t^2$. This non-locality in time reflects the discreteness of energy levels of reservoir oscillators and causes serious problems when evaluating the reduced dynamics. At low temperatures, the reduced quantum dynamics is always strongly retarded (non-Markovian) on time scales $\hbar\beta$, whence a simple time-local equation of motion does not generally exist. Equivalently, the idea of describing the time evolution of physical systems by means of equations of motion with suitable initial conditions is not directly applicable for quantum Brownian motion. It may hold approximately in certain limits such as very weak coupling (Breuer and Petruccione 2002) or, as we will see below, very strong dissipation.

The above procedure may also be understood from a different perspective. "What we observe as dissipation" in a system of interest is basically a consequence of our ignorance with respect to everything that surrounds this system. Dissipation is not inherent in nature, but rather follows from a concept that reduces the real world to a small part to be observed and a much larger part to be left alone. In the sequel, we will illustrate the consequences of this reduction, which are much more subtle than in the classical domain.

4.4 Specific Heat for a Brownian Particle

According to conventional thermodynamics, the specific heat (for fixed volume) is given by

$$C_v = \frac{\partial U}{\partial T},\tag{4.6}$$

where U denotes the internal energy of the system. Following classical concepts (Landau and Lifshitz 1958), the latter can be obtained either from the system energy (Hänggi et al. 2008)

$$\begin{aligned} U_E &= \langle H_S \rangle \\ &= \frac{\mathrm{Tr}\{\exp(-\beta H)H_S\}}{\mathrm{Tr}\{\exp(-\beta H)\}} \end{aligned}\tag{4.7}$$

or from the partition function

$$\begin{aligned} U_Z &= -\frac{\partial Z}{\partial \beta} \\ &= \frac{\mathrm{Tr}\{\exp(-\beta H)\}}{\mathrm{Tr}_R\{\exp(-\beta H_R)\}}. \end{aligned}\tag{4.8}$$

Here, we have used subscripts to distinguish between the two ways of obtaining the internal energy and thus the specific heat. Note that the partition function of the full compound system is defined with respect to the partition function of the bath alone. This is the only consistent way to introduce it, given that we average over the unobserved bath degrees of freedom. One easily realizes that the two routes may lead to quite different results, since

$$U_Z - U_E = \langle H_I \rangle + \langle H_R \rangle - \langle H_R \rangle_R.\tag{4.9}$$

Apart from the energy stored in the system-bath interaction, there also appears the difference in bath energies taken with respect to the full thermal distribution $\propto \exp(-\beta H)$ and the bath thermal distribution $\propto \exp(-\beta H_R)$. Classically, this difference vanishes due to the factorization of the thermal distribution. It may be negligible in the weak coupling regime where this factorization still holds approximately but certainly fails for stronger coupling and/or reservoirs with strongly non-Markovian behavior.

For a free quantum particle with so-called Drude damping, i.e., when $\gamma(t) = \gamma\omega_D \exp(-\omega_D t)$ or equivalently $J(\omega) = m\gamma\omega\omega_D/(\omega + \omega_D)$ with Drude frequency ω_D, one finds at low temperatures $T \to 0$ (Hänggi et al. 2008)

$$\frac{C_{v,E}}{k_B} \approx \frac{\pi k_B T}{3\hbar\gamma} \; , \; \frac{C_{v,Z}}{k_B} \approx \frac{\pi k_B T}{3\hbar\gamma}\left(1 - \frac{\gamma}{\omega_D}\right), \qquad (4.10)$$

in accordance with the third law of thermodynamics (vanishing specific heat for $T \to 0$). However, in contrast to $C_{v,E}$, the function $C_{v,Z}$ becomes negative for $\gamma/\omega_D > 1$, thus indicating a fundamental problem with this second way to obtain the specific heat. Apparently, the problem must be related to the definition (4.8) of the partition function of the reduced system. It does not exist in the high temperature regime or for Ohmic damping $\omega_D \to \infty$, and it is also absent for a harmonic system.

Any partition function can be expressed in terms of the density of states $\mu(E)$ of the system as $Z = \int_0^\infty dE\,\mu(E)\exp(-\beta E)$, which in turn allows us to retrieve $\mu(E)$ from a given partition function (Weiss 2008; Hänggi et al. 2008). Physically meaningful densities $\mu(E)$ must always be positive though. However, for reduced systems this is not the case in exactly those domains of parameter space where $C_{v,Z} < 0$. As a consequence, it only makes sense physically to start with the definition (4.7) for the specific heat, and this also implies that the partition function of a reduced quantum system does *not* play the same role as its counterpart in conventional classical thermodynamics.

This finding can once again be stated in the context of the measurement process in quantum physics: the expectation value of the system energy $\langle H_S \rangle$ is certainly experimentally accessible, while it seems completely unclear how to probe the partition function (4.8). One of its constituents, the partition function of the bare reservoir, cannot be measured as long as it is coupled to the system which, however, is not at the disposal of the experimentalist.

4.5 Roles Reversed: A Reservoir Dominates Coherent Dynamics

It is commonly expected that a noisy environment will tend to destroy quantum coherences in the system of interest and thus make it behave more classically. This gradual loss of quantumness has been of great interest recently because there has been a boost in activities to tailor atomic, molecular, and solid state structures with growing complexity and on growing length scales. A paradigmatic model is a two-state system interacting with a broadband heat bath of bosonic degrees of freedom (spin-boson model) which plays a fundamental role in a variety of applications (Weiss 2008; Breuer and Petruccione 2002; Leggett et al. 1987). Typically, at low temperatures and weak coupling, an initial non-equilibrium state evolves via damped coherent oscillations towards thermal equilibrium, while for stronger dissipation, relaxation occurs via an incoherent decay. This change from a quantum-type of dynamics to a classical-type with increasing dissipation at fixed temperature (or with increasing temperature at fixed friction) is often understood as a quantum to classical transition. It has thus been analyzed in great detail to tackle questions about the validity of quantum mechanics on macroscopic scales or the appearance of a classical world from microscopic quantum mechanics.

However, the picture described above does not always apply, as has been found only very recently (Kast and Ankerhold 2013a, b). In particular, at least for a specific class of reservoir spectral densities, so-called sub-Ohmic spectral densities, the situation is more complex, with domains in parameter space where the quantum-classical transition is completely absent even for very strong dissipation. This persistence of quantum coherence corresponds to a strong system-reservoir entanglement, such that the dynamical properties of the two-level system are dominated by properties of the bath.

A generic example of a two-level system is a double-well potential where two energetically degenerate minima are separated by a high potential barrier (Weiss 2008; Leggett et al. 1987). At very low temperatures, only the degenerate ground states $|L\rangle$ and $|R\rangle$ in the left and the right well, respectively, are relevant. They are coupled via quantum tunneling through the potential barrier with a coupling energy $\hbar\Delta$. Hence, the corresponding Hamiltonian for this two-level system follows as $H_S = (\hbar\Delta/2)(|L\rangle\langle R| + |R\rangle\langle L|)$ and the interaction with the bath is mediated via the operator $Q \rightarrow (|L\rangle\langle L| - |R\rangle\langle R|)$ in (4.4). In this way, the reservoir tends to localize the system in one of the ground states, while quantum coherence tends to delocalize it (superpositions of $|L\rangle$ and $|R\rangle$). The competition between the two processes leads to a complex dynamics for the populations $\langle L|\rho(t)|L\rangle = 1 - \langle R|\rho(t)|R\rangle$ and in most cases to damped oscillatory motion (coherent dynamics) for weak and monotonic decay (classical relaxation) for strong friction.

Sub-Ohmic reservoirs appear in many condensed phase systems where low frequency fluctuations are more abundant than in standard Ohmic heat baths. The corresponding spectral function

$$J_s(\omega) = 2\pi\alpha\omega_c^{1-s}\omega^s, \quad 0 < s < 1, \tag{4.11}$$

depends on the spectral exponent s, a coupling strength α, and a frequency scale ω_c. In the limit $s \rightarrow 1$, one recovers the standard Ohmic distribution. In thermal equilibrium and at zero temperature, a two-level system embedded in such an environment displays two "phases": a delocalized phase (quantum coherence prevails) for weaker friction and a localized phase (quantum non-locality destroyed) for stronger friction. The question then is: what does the relaxation dynamics towards these phases look like? The simple expectation is that the dynamics is quantum-like (oscillatory) in the former case and classical-like (monotonic decay) in the latter. That this is not always true is revealed by a numerical evaluation of (4.3).

As mentioned above, a treatment of the reduced quantum dynamics is a challenging task and, in fact, an issue of intense current research. While we refer to the literature [(Kast and Ankerhold 2013a, b) and references therein] for further details, we only mention that one powerful technique is based on the path integral

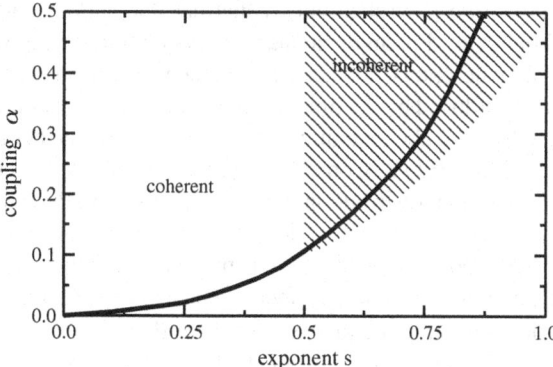

Fig. 4.1 Parameter space of a two-level system interacting with a sub-Ohmic reservoir with spectral exponent s and coupling strength α at $T = 0$. *Below* (*above*) the *black line*, for long times, the non-equilibrium dynamics approaches a thermal state with delocalized (localized) phase. This relaxation dynamics is incoherent only in the *shaded area* and coherent elsewhere, particularly, for $s < 1/2$ and also for strong friction

representation of (4.3) in combination with Monte Carlo algorithms. This numerical approach also allows one to access the strong friction regime in contrast to existing alternatives. As a result one gains a portrait in the parameter space of s, α (see Fig. 4.1) indicating the domains of coherent and incoherent non-equilibrium dynamics. Notably, for $0 < s < 1/2$, a transition from coherent to incoherent dynamics is absent even for strong coupling to the environment, and even though asymptotically the system approaches a thermal equilibrium with localized (classical-like) phase. It turns out that the frequency of this damped oscillatory population dynamics is given by $\Omega_s \approx 2\alpha\omega_c/s$ (for $s \ll 1$), with an effective damping rate $\gamma_0 \approx 2\alpha\omega_c$. Thus, the ratio $\Omega_s/\gamma_0 \approx 1/s \ll 1$, whence the system is strongly underdamped.

What is remarkable here is that these dynamical features of the two-level system are completely determined by properties of the reservoir. To leading order, the system energy scale $\hbar\Delta$ is negligible. In other words, on the one hand the system dynamics is slaved to the reservoir, and on the other hand the reservoir is no longer destructive but rather supports quantum coherent dynamics. This is only possible when system and bath are strongly entangled, which is indeed the case (Kast and Ankerhold 2013a). It is then at least questionable to consider the reservoir merely as a noisy background. Rather, in an experiment, one would access what are basically reservoir properties, whereas the two-level system acts only as a sort of mediator.

4.6 Emergence of Classicality in the Deep Quantum Regime

In conventional quantum thermodynamics one finds that, in the high temperature limit, classical thermodynamics is recovered. A simple example is a harmonic oscillator with frequency ω_0. If the thermal scale $k_B T$ far exceeds the energy level spacing $\hbar\omega_0$, quantization is washed out by thermal fluctuations and the oscillator displays classical behavior. For an open quantum system interacting with a real heat bath, an additional scale enters this scenario, namely, level broadening $\hbar\gamma$ (γ is a typical coupling rate) induced by the finite lifetime of energy eigenstates. In the weak coupling regime $\gamma\ll\omega_0$, the above argument then applies, and the classical domain is characterized by $\hbar\gamma\ll\hbar\omega_0\ll k_B T$. Consequently, the scale $\hbar\gamma$ does not play any role and the classical Boltzmann distribution $\exp(-\beta H_S)$ does not dependent on γ.

The same is true approximately at lower temperatures, where the canonical operator of the reduced system is given by that of the bare system, i.e., $\rho_\beta \propto \exp(-\beta H_S)$, as long as $\gamma\ll\omega_0$. In fact, even the reduced dynamics (4.2) can be cast into time-local evolution equations for the reduced density $\rho(t)$ if the time scale for bath-induced retardation $\hbar\beta$ is much shorter than the time scale for bath-induced relaxation $1/\gamma$. On a coarse-grained time scale, the reduced dynamics then appears to be Markovian. This domain

$$\frac{\hbar\gamma}{k_B T}\ll 1 \tag{4.12}$$

includes the weak friction regime $\gamma/\omega_0\ll 1$, in which so-called master equations are valid down to very low temperatures $\hbar\omega_0/k_B T\gg 1$ (Breuer and Petruccione 2002) (see Fig. 4.2). It also contains the classical regime of strong friction $\gamma/\omega_0\gg 1$ and very high temperatures $\omega_0\hbar/k_B T\ll 1$. Anyway, in the domain defined via (4.12), the

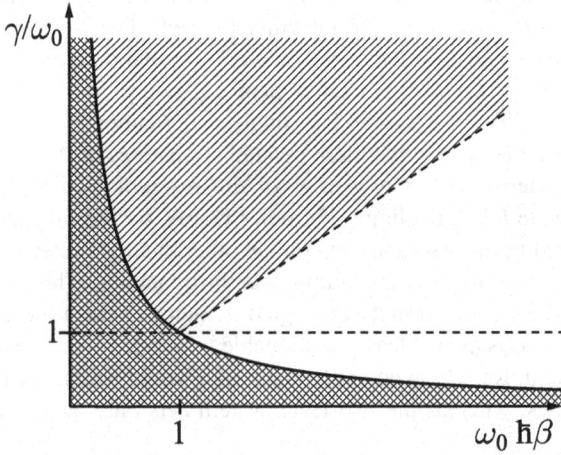

Fig. 4.2 Domains in parameter space for an open quantum system with typical energy level spacing $\hbar\omega_0$, interacting with a heat bath at reciprocal temperature β and with friction rate γ. *Below* the *black line* (*grid shaded*), the level broadening due to friction is negligible ($\gamma\hbar\beta\ll 1$), while *above* ($\gamma\hbar\beta\gg 1$), it must be treated non-perturbatively. The *dashed horizontal line* separates the domain of weak friction $\gamma/\omega_0\ll 1$ from the overdamped one $\gamma/\omega_0\gg 1$. At lower temperatures $\omega_0\hbar\beta > 1$, classicality appears in the line-shaded domain where a time scale separation $\gamma/\omega_0^2\gg\hbar\beta$ applies. See text for details

level broadening due to friction does not play a significant role and can be treated perturbatively.

The opposite is true for very strong dissipation $\gamma/\omega_0 \gg 1$ and low temperatures $\omega_0 \hbar/k_B T \gg 1$ (Ankerhold et al. 2001), so that

$$\frac{\hbar\gamma}{k_B T} \gg 1. \tag{4.13}$$

One might think that a simplification of the reduced dynamics is then not possible at all. However, it turns out that this is not quite true. Indeed, one must keep in mind that, in this regime, the relevant relaxation time scale for the reduced density is not given by $1/\gamma$, but rather by γ/ω_0^2. This follows directly from the classical dynamics of the harmonic oscillator, whose two characteristic roots are given by

$$\lambda_\pm = -\frac{\gamma}{2} \pm \sqrt{\frac{\gamma^2}{4} - \omega_0^2}. \tag{4.14}$$

In the overdamped limit, the fast frequency $\lambda_- \approx \gamma$ corresponds to the fast equilibration of momentum, while the slow frequency $\lambda_+ \approx \omega_0^2/\gamma$ describes the much slower relaxation of position. Accordingly, on a coarse-grained time scale, one may consider the momentum part of the reduced density to be equilibrated to the instantaneous position, so that the position part is the only relevant quantity for the dynamics [note that the system interacts with the bath in (4.4) via the position operator].

This separation of time scales is well known in classical physics as the overdamped or Smoluchowski limit (Risken 1984). There, it corresponds to a reduction of the Langevin Eq. (4.1) in which inertia effects are adiabatically eliminated. Equivalently, the position part $P(q, t)$ of the full phase space distribution of a one-dimensional particle with mass m moving in a potential $V(q)$ obeys the famous Smoluchowski equation (Smoluchowski 1906; Risken 1984), a time evolution equation in the form of a diffusion equation

$$\frac{\partial P(q,t)}{\partial t} = \frac{1}{\gamma m} \frac{\partial}{\partial q} \left[V'(q) + k_B T \frac{\partial}{\partial q} \right] P(q,t). \tag{4.15}$$

For quantum Brownian motion one can indeed show (Ankerhold et al. 2001; Ankerhold 2007; Maier and Ankerhold 2010) that a generalization of this equation, the so-called quantum Smoluchowski equation, follows from the reduced dynamics (4.2) in the overdamped regime where $\gamma/\omega_0^2 \gg \hbar\beta$ (ω_0 then refers to a typical energy scale of the system, see Fig. 4.2). The diffusion constant has to be replaced by a position-dependent diffusion coefficient $D(q) = k_B T/[1 - \beta V''(q)\Lambda]$ according to $k_B T \partial/\partial q \to \partial/\partial q D(q)$. In the high temperature limit, one recovers the result (4.15) from $\Lambda \to 0$, while in the domain (4.13), one has $\Lambda \propto \ln(\gamma\hbar\beta)/\gamma$. Then, the dependence on the bath coupling γ and Planck's constant $\hbar$ appears in a highly non-perturbative way and may substantially influence the dynamics. However, the equation which governs this dynamics has a basically classical structure.

One may also understand the origin of an $\hbar$-dependent diffusion coefficient from Heisenberg's uncertainty relation. In the strong coupling regime, the reservoir squeezes fluctuations in position to the extent that the position distribution obeys a semi-classical type of equation of motion. This in turn requires momentum fluctuations to be strongly enhanced, which can indeed be shown. The quantum parameter Λ is reminiscent of this interdependence between position and momentum fluctuations. What is interesting experimentally, is the fact that, according to (4.13), quantum fluctuations may already appear at very high temperatures ($\hbar\omega_0 \ll k_B T$) if friction is sufficiently strong as, for example, in biological structures.

4.7 Summary and Conclusion

In this contribution we have shed some light on the subtleties that are associated with the description of open quantum systems. In contrast to the procedure in classical physics, quantum Brownian motion can only be consistently formulated when given a Hamiltonian of the full system. In a first step, this requires one to identify the relevant system part and its irrelevant surroundings, a choice which is not unique and depends on the focus of the observer. One then implements a reduction, keeping only the effective impact of the environment by assuming that it constitutes a heat bath. This has at least two substantial advantages: (i) a microscopic description of the actual reservoir is not necessary and (ii) this modeling provides a very general framework, applicable to a broad class of physical situations. In fact, it has turned out to be the most powerful approach we have for understanding experimental data for dissipative quantum dynamics. However, the price to pay for this reductionism is that, on the one hand, the dynamics cannot generally be cast into the form of simple time evolution equations, and on the other, conventional concepts and expectations must be treated with great caution.

The fundamental process is once again the non-locality of quantum mechanics, which in this context may lead to a "blurring" of what is taken to be the system and what is taken to be its surroundings. This entanglement becomes particularly serious when the interaction between the system and the heat bath is no longer weak. We have discussed here one example from thermodynamics and two examples from non-equilibrium dynamics. With respect to the first, it was shown that the partition function of a reduced system is not a proper partition function in the conventional sense and so cannot always be used to derive thermodynamic quantities. In the second and the third example, our naive conception of the division of the world into a classical realm and a quantum realm has been challenged. There may be emergence, with the consequence that quantum mechanics may survive even for strong friction and classicality may be found even in the deep quantum domain.

References

Ankerhold, J.: Quantum Tunneling in Complex Systems. Springer Tracts in Modern Physics, vol. 224. Springer, Berlin (2007)

Ankerhold, J., Pechukas, P., Grabert, H.: Phys. Rev. Lett. **87**, 086802 (2001)

Blum, K.: Density Matrix Theory and Applications. Plenum Press, New York (1981)

Breuer, H.P., Petruccione, F.: The Theory of Open Quantum Systems. Oxford University Press, Oxford (2002)

Einstein, A.: Ann. Phys. (Leipzig) **17**, 549 (1905)

Gardiner, C.W., Zoller, P.: Quantum Noise. Springer, Berlin (2004)

Hänggi, P., Ingold, G.-L., Talkner, P.: New J. Phys. **10**, 115008 (2008)

Kast, D., Ankerhold, J.: Phys. Rev. Lett. **110**, 010402 (2013a)

Kast, D., Ankerhold, J.: Phys. Rev. B **87**, 134301 (2013b)

Landau, L.D., Lifshitz, E.M.: Statistical Physics. Pergamon, Oxford (1958)

Leggett, A.J., Charkarvarty, S., Dorsey, A.T., Fisher, M.P.A., Garg, A., Zwerger, W.: Rev. Mod. Phys. **59**, 1 (1987)

Maier, S., Ankerhold, J.: Phys. Rev. E **81**, 021107 (2010)

Redfield, A.G.: IBM J. Res. Dev. **1**, 19 (1957)

Risken, H.: The Fokker-Planck Equation. Springer, Berlin (1984)

Smoluchowski, M.: Ann. Phys. (Leipzig) **21**, 772 (1906)

Wangsness, R.K., Bloch, F.: Phys. Rev. **89**, 728 (1953)

Weiss, U.: Quantum Dissipative Systems. World Scientific, Singapore (2008)

Chapter 5
Explanation Via Micro-reduction: On the Role of Scale Separation for Quantitative Modelling

Rafaela Hillerbrand

5.1 Introduction

In many areas of philosophy, microphysicalism seems an undisputed dogma. As Kim (1984, p. 100) puts it: "ultimately the world—at least, the physical world—is the way it is because the micro-world is the way it is [...]." But not only on the ontological, but also on the epistemic level, there is a clear preference for a micro-level description over a macro-level one. At least within the physical sciences, the governing laws on the micro-level are often seen as somewhat 'simpler' than the macro-level behavior: here universal laws can be applied to the movement of individual atoms or molecules, while the macro-level seems to behave in a rather intricate way.

Condensed matter physics and particularly solid state physics as its largest branch provide examples at hand: equilibrium and non-equilibrium statistical mechanics together with electromagnetism and quantum mechanics are used to model the interactions between individual atoms or molecules in order to derive properties of macro-phenomena like phase transitions, e.g. the melting of water or the onset of laser activity. More generally, the whole study of complex systems—for example, within the complexity theory, various theories of self-organization such as Haken's synergetics, or Prigogine's non-equilibrium thermodynamics—builds on the reductionist paradigm.

In all these cases, the ontological reduction is not disputed. A solid body is made up from atoms; a complex system is even defined by its constituents: it is made up

R. Hillerbrand (✉)
TPM-Faculty of Technology, Policy and Management TU Delft, Jaffalaan 5, 2628BX Delft, The Netherlands
e-mail: r.c.hillerbrand@tudelft.nl

© Springer-Verlag Berlin Heidelberg 2015
B. Falkenburg and M. Morrison (eds.), *Why More Is Different*,
The Frontiers Collection, DOI 10.1007/978-3-662-43911-1_5

of multiple interconnected parts. As indicated in the above citation by Kim, this often leads philosophers to claim an ontological priority for the micro level. This priority has recently been called into question by Hüttemann (2004). But what does this imply about a possible explanatory or epistemic priority of the micro over the macro description?

This paper leaves questions regarding the ontological priority aside and focusses on scientific models in which the ontological micro-reduction is not under dispute. I address whether and how an *ontological* reduction may entail an *explanatory* reduction. Hence, this paper is not about explanatory or epistemic reduction in its most general sense, but zooms in on a specific type of reductionist explanation, or rather prediction, in which quantitative information about macro-level quantities such as temperature or pressure are derived from the micro-level description. It will be shown that this is the case only for certain types of micro-reductionist models, namely those exhibiting scale separation. A separation of scales is present in the mathematical micro-model when the relevant scales differ by an order of magnitude.[1] To illustrate what scale separation amounts to, consider as a very simple example, discussed in more detail later in the paper, the motion of the Earth around the Sun. This can be treated as a two-body problem as the forces exerted on the Earth by all other celestial bodies are *smaller by orders of magnitude* than the force exerted by the Sun. We have a separation of scales here where the relevant scale is force or energy. Relevant scales may also be time, length, or others.

The epistemic pluralism I aim to defend on the basis of this analysis is not just another example of a non-reductionist physicalism. Rather it suggests how to draw a line between emergence and successful micro-reduction in terms of criteria the micro-formulation has to fulfill. It is argued that one should consider scale separation as a specific class of reduction. It will be shown that the identified criterion, scale separation, is of much broader relevance than simply the analysis of condensed matter systems. Hence in this sense, these fields do not occupy a special position concerning their methodology. Scale separation as a criterion for epistemically successful micro-reduction has important implications, particularly for the computational sciences where micro-reductionist models are commonly used.

The following Sect. 5.2 introduces the type of reductionist models this paper is concerned with, i.e. mathematically formulated micro-reductionist models in which the ontological reduction is not under dispute. The criteria of successful reduction will be detailed in terms of what is referred to as 'generalized state variables' of the macro system. Section 5.3 discusses two similar mathematical models in non-equilibrium physics that are formulated on a micro level to predict certain macro quantities, namely: the semi-classical laser theory in solid-state physics and current models of turbulent flows in fluid dynamic turbulence. It is argued that the relevant

---

[1] The term 'scale separation' is borrowed from the theory of critical phenomena, but applied more generally in this paper.

difference in the mathematical models is that while the former exhibits scale separation, the latter does not. The absence of scale-separation, as will be shown, renders the contemporary micro-reductionist approaches to turbulence unsuccessful, while the semi-classical laser theory allows for quantitative predictions of certain macro variables. The laser is chosen as a case study because despite being a system far from equilibrium it allows for quantitative forecasts. Section 5.4 puts the debate in a broader context, showing that a separation of scales is not only relevant for complex systems, but also for fundamental science. Section 5.5 generalizes the concept to local scale separation, i.e. cases where more than one scale dominates. Section 5.6 discusses the importance and implications the special role of scale-separation in micro-reduction has for various questions within the philosophy of science, such as emergent properties and the use of computer simulations in the sciences.

5.2 Explanation and Reduction

Solid state physics derives the macro-properties of solids from the material's properties on the scale of its atoms and molecules and thus seems a paradigm example of a (interlevel or synchronic) reductionist approach within the sciences. Within the philosophy of science, this claim is, however, controversial: as pointed out by Stöckler (1991) well-established tools within the field, such as the so-called 'slaving principle' of Haken's theory of synergetics, are seen by proponents of reductionist views as clear-cut examples for reduction, while anti-reductionists use the very same examples to support their views.

In order to make sense of those contradictory views, we need to analyze in more detail what reduction refers to: speaking of a part-whole relation between the atoms and the solid body refers to ontological statements, while the relation between the macroscopic description of the solid body that is derived from the evolution equation of its atoms and molecules refers to the epistemic description. In Sect. 5.2.1 I want to follow Hoyningen-Huene (1989, 1992, 2007) and distinguish methodological, ontological, epistemic, and explanatory reduction.[2] The criteria of a 'successful' micro reduction will be spelled out in Sect. 5.2.2: I focus on the prediction of certain macro-features of the system in quantitative terms. Throughout this paper the focus is on *micro*-reductions as a part-whole relation between reduced r (macro) and reducing level R (micro), $r \rightarrow R$; however, the reduced and reducing descriptions do not need to correspond to micro and macro*scopic* descriptions.

---

[2] Hoyningen-Huene distinguishes further types of reduction which for the purpose of this paper that focuses on mathematically formulated models and theories are not of relevance.

5.2.1 Types of Reduction

Methodological reduction concerns the methodologies used to study the reduced and reducing phenomena. For example a methodological reductionism within biology assumes that biological phenomena can only be adequately studied with the methods of chemistry or physics.

Ontological reduction is concerned with questions about whether the phenomena examined on the reduced r and the reducing level R differ *as regards their substantiality*. The classical example, the reduction of the phenomena of thermodynamics to those of statistical mechanics, is clearly a case of ontological reduction: a thermodynamic system such as a gas in a macroscopic container is made up of its molecules or atoms. Exactly the same is true for a solid body that reduces ontologically to its constituents.

Such an ontological reduction can be distinguished from epistemic reduction. The latter considers the question as to whether knowledge of the phenomena on the reduced level r can be ascribed to knowledge of the R-phenomena. As in the thermodynamic example, the knowledge of the R-phenomena are typically formulated in terms of (natural) laws. Together with certain other specifications, like bridge rules that connect the vocabulary of the reduced and reducing description, the R-laws can be used to derive the laws of r. Along these lines it is often said that the *laws* of thermodynamics (r) reduce to those of statistical mechanics (R). Note however that unlike ontological reduction, even this example of statistical mechanics and thermodynamics is not an undisputed example of epistemic or explanatory reduction.

From this epistemic reduction we may also ask whether the r-phenomena can be *explained* with the help of the resources of the reducing level R. This is referred to as explanatory reduction. Note that when we assume, like Nagel (1961) did in his seminal writing on reduction, a deductive-nomological model of explanation, then explanation coincides with deduction from natural laws, and hence explanatory reduction coincides with epistemic reduction. However not all types of explanations are of the deductive-nomological type (see Wimsatt 1976).

Note that even in cases where law-like explanations exist on the level of the reducing theory, the correspondence rules may be in need of explanation. Hence, despite a successful epistemic reduction, we may not accept it as an explanation and hence as an explanatory successful reduction. Hoyningen-Huene mentions the example of epistemic reduction of psychological states to neuronal phenomena. The correspondence rule relates certain psychological sensations like the sentiment of red to certain neural states. Why this sentiment relates to a certain neural state, remains in need of an explanation.[3]

---

[3] It should be noted that when paradigm cases of reduction like thermodynamics are contested it is often the bridge rules that are problematic.

As Hoyningen-Huene points out, a generic explication of explanatory reduction needs to remain vague because it needs to leave room for various types of explanations. For the remainder of this paper, the focus is on a specific type of explanatory reduction that is detailed in the following Sect. 5.2.2. Though the analysis remains in the realm of mathematical modeling, I want to refer to the reduction as 'explanatory' instead of 'epistemic', because in many areas of science it is not so much laws of nature or theories that play a central role, but models (see Morgan and Morrison 1999). Nonetheless, theories may play a central role in deriving the models. I will argue that the micro-level R-description needs to entail certain features in order for an undisputed ontological reduction to lead to a successful explanatory reduction.

5.2.2 Quantitative Predictions and Generalized State Variables

Complexity theory provides an excellent example of a methodological reduction: chaos theory or synergetics, first developed in the context of physical systems, are now also applied to biological or economic systems (e.g. Haken 2004). Ontological reduction is not under dispute here; my concern is with uncontested examples of ontological micro-reduction. For these I will investigate in more detail the epistemic or explanatory aspects of the reduction. My main thesis is that ontological reduction does not suffice for explanatory reduction, not even in those cases where a mathematical formulation of the micro-level constituents can be given. In this Sect. 5.2.2 I explicate in more detail what I characterise as a successful explanatory reduction, thereby distinguishing an important subclass of reduction that has not yet received attention within the philosophy of science.

A scientific model commonly does not represent all aspects of the phenomena under consideration. Within the literature on models and representation, often the term 'target system' is used to denote that it is specific aspects that our scientific modeling focuses on. Consider a simple example of a concrete model, namely a scale model of a car in a wind tunnel. This model does not aim to represent all aspects of the car on the road. Rather the target system consists of a narrow set of some of the car's aspects, namely the fluid dynamical characteristics of the car, such as its drag. The model does not adequately represent and does not aim to adequately represent other features of the car, such as its driving characteristics. This also holds true for abstract models and more generally for any theoretical description. This seemingly obvious feature of scientific investigation has important consequences. For our discussion of explanatory reduction, it implies that we need to take a closer look at which aspects of the r-phenomena we aim to predict from the micro description R.

In what follows I want to focus on one specific aspect that is important in a large number of the (applied) sciences, namely the prediction of certain quantitative aspects of the macro system such as the temperature and pressure at which a solid body melts. This information is quite distinct from many investigations within complexity theory. Here, the analysis on the microlevel often gives us information on the stability of the system as a whole. Chaos theory allows us to predict whether a system is in a chaotic regime, i.e. whether the system reacts sensitively to only minute changes in certain parameters or variables. Here, for example, the Lyapunov exponents provide us with insights into the stability of the macro system. The first Lyapunov exponent characterizes the rate at which two infinitesimal trajectories in phase space separate. The conditions under which this information can be derived are well understood and detailed in Oseledets' Multiplicative Ergodic theorem (Oseledets 1968).

The information about the Lyapunov exponent is quantitative in nature insofar as the exponents have a certain numerical value. However, this quantitative information on the reduced level only translates into a qualitative statement about the macro system's stability, not quantitative information about its actual state. My concern here is much more specific: it focuses on when a micro-reductionist model can be expected to yield information of the system's macro properties that are described in quantitative terms. I refer to those quantities that are related to the actual state of the system as 'generalized state variables'. The term 'state variable' is borrowed from thermodynamics and denotes variables that have a unique value in a well-defined macro state. *Generalized* state variable indicates that unlike thermodynamics, the value may depend on the history of the system.[4] So the question to be addressed in the following is when a micro-reductionistic model can yield quantitative information about the generalized state variables of the macro system?

5.3 Predicting Complex Systems

I will contend in the following that a separation of scales is the requirement for successful micro-reduction for complex systems with a large number of interacting degrees of freedom on the microlevel. Therefore I argue that one should distinguish a further class of micro models, namely those that exhibit scale separation. Depending on the phenomenon under consideration and on the chosen description, these scales can be time, length, energy, etc.

A very simple example, the movement of the Earth around the Sun modeled as a two body problem, is used to illustrate the concept of scale separation in Sect. 5.3.1. Arguments for scale separation as a condition for quantitative predictions on the generalized state variables are developed by two case studies: I compare two similar,

---

[4] This feature will only become relevant in the latter sections of this paper, when I briefly touch on the applicability of my approach beyond many-body systems.

highly complex, non-equilibrium dynamical systems, namely the laser (Sect. 5.3.2) and fluid dynamic turbulence (Sect. 5.3.3). I claim that while the former exhibits scale separation, the latter does not and this will be identified as the reason for a lack of quantitative information regarding the macro level quantities in the latter case.

5.3.1 Scale Separation in a Nutshell

Before beginning my analysis, let me expand a little on a common answer to the question of when mathematical models are able to lend themselves to quantitative forecasts. Recall the oft-cited derivation of the model of the Earth's motion around the Sun from Newton's theory of gravitation and the canonical formulation of classical mechanics in terms of Newton's laws. This may be seen as a micro-reductionist approach in the sense introduced above: The goal is to derive information on the macro level, e.g. on the length of the day-night cycle, from the reduced micro-level that consists of planetary motions. The Earth-Sun motion clearly is a model allowing for quantitative predictions. The common answer as to why this is so is that the evolution equations are integrable. As early as 1893 Poincaré noted that the evolution equations of more than two bodies that interact gravitationally are no longer integrable and might yield chaotic motion (Poincaré 1893, pp. 23–61). One common answer to the question as to whether or not quantitative forecasts of the state variables are feasible thus seems to depend on the complexity of the system: complex systems, i.e. systems with a large number of degrees of freedom (in this case larger than two) that are coupled to each other via feedbacks, and hence exhibit nonlinear evolution equations of their variables, resist quantitative predictions.

To deal with these complex systems, we have to rely on complexity theory or chaos theory in order to determine information about the stability of the systems and others dynamical properties. But as noted above, we are often not interested in this information on the system's stability alone, but want quantitative information on generalized state variables. Although, strictly speaking, the motion of the Earth around the Sun is a multi-body problem it is possible to reduce its motion to an effective two-body problem. So the important question is when the two-body approximation is adequate. And this is not addressed by Poincare's answer. Note that the model of planetary motion goes beyond Newton's theory insofar as it entails additional assumptions—such as neglecting the internal structure of the planets, specific values for the bodies' masses, or neglecting other bodies such as other planets or meteorites. With the model comes a tacit knowledge as to why these approximations are good approximations: the motion of the Earth can well be described by its motion around the Sun only because the forces exerted on the Earth by other bodies, even by the large planets like Saturn and Jupiter, are smaller *by orders of magnitude* than the force exerted by the Sun. If there were large scale objects, say, of the size of Jupiter or Saturn that orbit around some far away center, but come very close to the orbit of the Earth from time to time, then the reduced description in terms of two bodies would break down.

Thus the model of the Earth-Sun motion works because the forces exerted on the Earth by all other bodies, planets asteroids etc. are smaller by orders of magnitude than the force exerted on the Earth by the presence of the Sun. With information about the distance between the relevant bodies, the forces can be directly translated in terms of work or energy. In other words, the micro model exhibits a separation of the requisite energy scales. Generally speaking, scale separation may refer to any scale like energy, time, length, or others, depending on the formulation of the micro model.

5.3.2 Lasers

The so-called "semi-classical laser theory", originally developed by H. Haken in 1962, is a very successful application of the micro reductionist paradigm detailed in Sect. 5.2. The properties of the macro-level laser are derived from a quantum mechanical description of the matter in the laser cavity and its interaction with light which is treated classically; the propagation of light is derived from classical electromagnetism. As Haken (2004, p. 229) puts it "[t]he laser is nowadays one of the best understood many-body problems." As I will show in this section, the key feature of the micro description that allows for a successful reduction is the display of scale separation.

A laser is a specific lamp capable of emitting coherent light. The acronym 'laser', light amplification by stimulated emission of radiation, explicates the laser's basic principle:

- The atoms in the laser cavity (these may be a solid, liquid, or gas) are excited by an external energy source. This process is called pumping. It is schematically depicted in Fig. 5.1 for the most simple case of a two-mode laser: atoms in an excited state. On relaxing back to their ground state, the atoms emit light in an incoherent way, just like any ordinary lamp.[5]
- When the number of excited atoms is above a critical value, laser activity begins: an incident light beam no longer becomes damped exponentially due to the interaction with the atoms, rather, the atoms organize their relaxation to the ground state in such a way that they emit the very same light in terms of frequency and direction as the incident light beam. This is depicted in Fig. 5.1: the initial light beam gets multiply amplified as the atoms de-excite, thereby emitting light that in phase, polarization, wavelength, and direction of propagation is the same as the incident light wave—the characteristic laser light. One speaks of this process also as self-organization of the atoms in the laser medium.

---

[5] Note that this phase transition is not adequately described within the semiclassical approach as the emission of normal light due to spontaneous emission necessitates a quantum mechanical treatment of light.

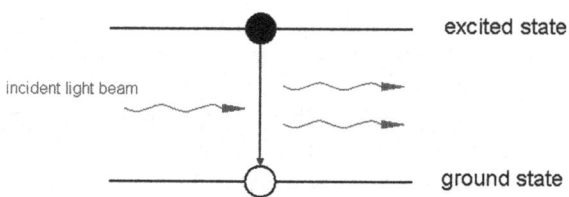

Fig. 5.1 Schematic sketch of a laser cavity. The incident light beam triggers the atom in the excited state (*black circle*) to relax to its ground state (*white circle*), thereby emitting light with the same phase, polarization, wavelength, and direction of propagation as the incident light wave

The laser is discussed here because it is a fairly generic system: first, it is a system that features a large number of interacting degrees of freedom. The incoming light beam interacts with the excited atoms of the laser medium in the cavity; the number of atoms involved—*all* the atoms in the laser cavity—is of the order of the Avogadro number, i.e. of the order of 10^{23}. Secondly, it is a system far from equilibrium in that the 'pumping' energy must be supplied to keep the atoms in the excited state. Thirdly, the system is dissipative as it constantly emits light.

The light is described classically in terms of the electric field $E(x,t)$ which obeys Maxwell's equations, and decomposed into its modes $E_k(t)$, where $k = 1, 2, \ldots$ labels the respective cavity mode. The semi-classical laser equations describe the propagation of each mode in the cavity as an exponentially damped wave propagation through the cavity, superimposed by a coupling of the light field to the dipole moments $\alpha_i(t)$ of the atoms and stochastic forces that incorporate the unavoidable fluctuations when dissipation is present. The dipole moment of atom i is itself not static, but changes with the incident light and the number of electrons in an excited state. To be more precise, the evolution equation of $\alpha_i(t)$ couples to the inversion $\sigma_i(t)$ which is itself not static, but dependent on the light field E. Here the index i labels the atoms and runs from 1 to $O(10^{23})$.

This leaves us with the following evolution equations:

$$
\begin{aligned}
\dot{E}_\lambda &= (-i\omega_\lambda - \kappa_\lambda)E_\lambda - i\sum_k g_{k\lambda}\alpha_k + F_\lambda(t), \\
\dot{\alpha}_k &= (-i\nu_k - \gamma)\alpha_k + i\sum_\lambda g_{k\lambda}^* E_\lambda \sigma_k + \Gamma_\lambda(t), \\
\dot{\sigma}_k &= \bar{\gamma}(d_0 - \sigma_k) + 2i\sum_\lambda \left(g_{k\lambda}\alpha_k E_\lambda^\dagger + c.c.\right) + \Gamma_{\sigma,\lambda}(t),
\end{aligned}
\tag{5.1}
$$

where $F_\lambda(t)$, $\Gamma_\lambda(t)$ and $\Gamma_{\sigma,\lambda}(t)$ denote stochastic forces.

The details and a mathematical treatment can be found in Haken (2004, pp. 230–240). For our purposes it suffices to note that as it stands, the microscopic description of the matter-light interaction is too complex to be solved even with the help of the largest computers available today or in the foreseeable future: for every mode k of the electric field, we have in Eq. (5.1) a system of the order of 10^{23} evolution equations, namely three coupled and thus nonlinear evolution equations for each atom of the cavity.

The methodology known as the 'slaving principle' reflects the physical process at the onset of the laser activity and allows us to drastically simplify the micro-reductionist description. Generally we will expect the amplitude of the electric field E to perform an exponentially damped oscillation. However, when the inversion σ_i and thus the number of atoms in excited states increases, the system suddenly becomes unstable and laser activity begins, i.e. at some point coherent laser light will be emitted. This means that the light amplitude becomes virtually undamped and thus the internal relaxation time of the field is very long. In fact, when this happens the amplitude of the electric field is by far the slowest motion of the system. Hence the atoms, which move very fast compared to the electric field, follow the motion of the light beam almost immediately. One says that the electric field *enslaves* all other degrees of freedom, i.e. the inversion and the dipole moment of all the involved atoms. Hence, the term 'slaving principle' describes a method for reducing the number of effective degrees of freedom. This allows elimination of the variables describing the atoms, i.e. the dipole moments α_i and the inversion σ_i. From the set of laser equations above, one then obtains a closed evolution equation for the electric field E that can be solved.

The crucial feature of the model that enables the elimination of the atomic variables is that the relaxation time of the atomic dipole moment is *smaller by far* than the relaxation time of the electric field. Expressed differently: the relevant time scales in this model separate. Consequently, the onset of laser activity at a fixed energy level can be predicted from the micro-reductionist model that describes the atoms and the light with the help of quantum mechanics and electromagnetism.

5.3.3 Fluid Dynamic Turbulence

In order to stress that indeed the separation of (time) scales is the relevant feature that allows for a successful micro-reduction in the sense defined in Sect. 5.2, I contrast the semi-classical laser theory with the description of fluid dynamic turbulence. The two case studies are similar in many respects: both are systems far from equilibrium, both are distinguished by a large number of degrees of freedom, and the interacting degrees of freedom give rise to nonlinear evolution equations. Moreover, in the current description, both systems do not display an obviously small parameter that allows for a perturbative treatment.

The equivalent equations for fluid motion are the Navier-Stokes equations which in the absence of external forcing, are:

$$\frac{DU_i}{Dt} = -\frac{\partial}{\partial x_i}p + \frac{1}{Re}\frac{1}{\partial_j\partial_j}v_i \, , \tag{5.2}$$

where U_i denotes the velocity field, Re is a dimensional number that characterizes the state of the flow (high Re corresponds to turbulent flows), and $D/Dt = \partial/\partial t + U_i \partial/\partial x_i$ the substantial derivative. Unlike the laser Eq. (5.1), these are not direct descriptions on the molecular level, but mesoscopic equations that describe the movement of a hypothetical mesoscopic fluid particle, maybe best understood as a small parcel of fluid traveling within the main flow. The Navier-Stokes equations are the micro-reductionist descriptions I want to focus on and from which one attempts to derive the macro-phenomena of turbulence by the evolution equations governing its constituents.[6]

For a fully developed turbulent flow, the Reynolds number Re is too large to solve these equations analytically or numerically.[7] This is similar to the semi-classical laser theory. In the latter case the slaving principle provided us with a method for reducing the number of degrees of freedom, something this is not yet possible for fully developed turbulent flows.

While there exist fruitful theoretic approaches to turbulence that can successfully model various aspects of it, these remain vastly unconnected to the underlying equations of motion (5.2) that express the micro-reductionist description. Theoretical approaches are based on scaling arguments, i.e. on dimensional analysis (Frisch 1995), or approach the issue with the help of toy models: Recognizing certain properties of Eq. (5.2) as pivotal in the behavior of turbulence, one identifies simpler models that share those properties and that can be solved analytically or numerically (Pope 2000). Various models of turbulence in the applied and engineering sciences make use of simplified heuristics (Pope 2000). In Sect. 5.5, I will come back to one of these approaches, the so-called 'large eddy simulations' that underly many numerical simulations of turbulence, where local scale separation is discussed as a weaker requirement than scale separation.

Despite the success of these theoretical approaches to turbulence, it remains an unsolved puzzle as to how to connect the toy models or the scaling laws to the underlying laws of fluid motion (Eq. 5.2). Common to all these approaches to turbulence is that they are models on the macro-level and are not derived from the micro-level model expressed by the Navier-Stokes equations. The reductive approach in the case of turbulence has not yet been successful, which leaves turbulence, at least for the time being, as "the last unsolved problem in classical physics" (Falkovich and Sreenivasan 2006). In order to analyse as to why this is so, let us take a closer look at the physics of a turbulent flow.

---

[6] The Navier-Stokes equations can be derived from the micro-level description given by Liouville's equation that gives the time evolution of the phase space distribution function. The derivation of the Navier-Stokes equation from even more fundamental laws, is, however, not the concern of this paper.

[7] There have been, however, recent advances in direct numerical simulations of the Navier-Stokes equations in the turbulent regime. However, the investigated Reynolds number today are below those investigated in laboratory experiments and it takes weeks or month of CPU. Also boundaries pose a severe challenge to the direct numerical simulations of the Navier-Stokes equations (see Pope 2000).

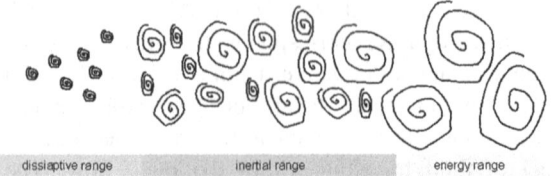

dissiaptive range   inertial range   energy range

Fig. 5.2 Fluid dynamic turbulence: large eddies of the size of the flow geometry are created and transported by the mean flow (here from *right* to *left*), thereby breaking up into smaller eddies. Within the smallest eddies (*left*), the turbulent energy is dissipated into mean energy of the flow. Note that within the so-called 'inertial range', eddies of various sizes coexist

Turbulence is characterized by a large number of active length or time scales which can be heuristically associated to eddies of various sizes and their turnover time, respectively (see Fig. 5.2). The simplest heuristic picture of turbulence is that eddies of the size of the flow geometry are generated: these eddies are then transported along by the mean flow, thereby breaking up into smaller and smaller eddies. This so-called 'eddy cascade' ends in the smallest eddies in which the energy of the turbulent motion is again dissipated into energy of the mean flow. It is the existence of such a range of active scales that spoils the success of the micro-reductionistic approach. When we focus on the eddies' turn-over time the analogy, or rather the discrepancy, with the laser becomes most striking: for the laser, at the point of the phase transition there is one time-scale that slaves all other degrees of freedom (as these are smaller by orders of magnitude), but there exists a large number of active time-scales in a turbulent flow.

To summarize, two complex, non-equilibrium systems were addressed: The laser and fluid dynamic turbulence. The mathematical formalization in both descriptions is fairly similar, both are problems within complex systems theory. However, while the first problem allows one to derive macroscopic quantities of interest from the micro-model, for turbulence this is not possible. The reason for this is a lack of separation of time and length scales because in turbulent flows eddies of various sizes and various turn-over times are active.

5.4 Scale Separation, Methodological Unification, and Micro-Reduction

A separation of scales is featured in many complex systems. Indeed, the term 'slaving principle' was originally coined within the semiclassical description of the laser, but is now widely applied in various areas of complex systems, chaos or catastrophe theory (e.g. Zaikin and Zhabotinsky 1970). The methodology coincides with what is sometimes referred to as 'adiabatic elimination'. The slaving principle has been also applied successfully to mathematical models within biology and the social sciences (Haken 2004). We may see its use as an example of a successful methodological reduction. However, in order to argue that micro-reduction with scale separation constitutes an important class of explanatory reduction, one must ask whether this is a useful concept only within the analysis of complex systems. After all, it is often argued that condensed matter physics is special within the physical sciences in that it relates to statistical mechanics or thermodynamics and cannot be fully reduced to the theories of the four fundamental forces, i.e. electromagnetic, gravitational, strong, and weak interaction. In the following Sect. 5.4.1 I therefore want to show that scale separation is also an essential characteristic of successful micro-reductions in other areas of physics. The section shows how the idea of a separation of scales can also accommodate scale-free systems like critical systems within equilibrium statistical mechanics.

5.4.1 Fundamental Laws: Field Theories and Scale Separation

We can show how scale separation is a key issue for deriving quantitative predictions about macro phenomena from a micro-reductionistic approach even in areas with an elaborate theoretical framework such as quantum chromodynamics (QCD) and quantum electrodynamics (QED). Despite their differences, these two theories are technically on a similar advanced level: Both are gauge theories, both are known to be renormalizable, we understand how a non-Abelian gauge theory like QCD differs from the Abelian theory of QED, and so forth. For QED we can derive good quantitative predictions. Indeed, perturbative models deriving from quantum field theory constitute the basis for large parts of quantum chemistry. However, for QCD, this is only possible at small distances or, equivalently, in the high energy regime. This is due to the fact that, while the coupling constant α, i.e. a scalar-valued quantity determining the strength of the interaction, is small for electromagnetic interactions, in fact, $\alpha_{QED} \approx 1/137 \ll 1$, the coupling constant α_{QCD} for strong interactions is large, except for short distances or high energies. For example, it is of the order of 1 for distances of the order of the nucleon size. Hence, speaking somewhat loosely, since field theories do not contain the term 'force', for strongly interacting objects, the different forces can no longer be separated from each other.

It is the smallness of the coupling constant in QED that allows a non-field theoretical approach to, for example, the hydrogen atom: fine and hyperfine structure are nicely visible in higher order perturbation theory. The heuristic picture behind the physics of strong interactions, as suggested by our account of scale separation, is that here the (chromo-electric) fine structure and hyperfine structure corrections are of the order of the unperturbed system, i.e. the system lacks a separation of the relevant scales. More generally as regards modern field theories, one may see the success of so-called 'effective field theories' in contemporary theoretical particle physics as an application of the principle of scale separation (Wells 2013).

5.4.2 Critical Phenomena

Section 5.3 introduced the concept of scale separation by means of the laser which involves a non-equilibrium phase transition. Also, equilibrium phase transitions as described by the theory of critical phenomena can be understood in terms of a separation of scales, in this case length scales. I focus on this below because it shows how scale-free systems may originate from scale separation.

A typical example of a continuous equilibrium phase transition is a ferromagnet that passes from non-magnetic behavior at high temperatures to magnetic behavior when the temperature decreases below the critical Curie temperature. While the description of such equilibrium statistical mechanical systems is formulated on a micro scale of the order of the atomic distance within the ferromagnet, one is actually interested in the macroscopic transition from a nonmagnetic to a magnetic phase, for which many of the microscopic details are irrelevant.[8] Within critical theory, one speaks here of 'universality classes' when different microscopic realizations yield the same macroscopic properties.

For a long time such phase transitions remained an unsolved problem due to their remarkable features, in particular the universal behaviour of many system at the critical point, i.e. at the phase transition. Classical statistical mechanics simply fails to account for these phenomena due to density fluctuations. Let us take a closer look at the physics at the point of the phase transition.

With the help of the renormalization group we can integrate out the microscopic degrees of freedom. This procedure yields a course-grained description similar to that encountered in deriving the Navier-Stokes equations. The central point in the Ising model, a crude model of a magnet, is an invariance at the critical point, namely: the free energy is invariant under coarse-graining. This is so because the system is self-similar at the point where the phase transition occurs: no characteristic length scale(s) exists as one of the correlation lengths, i.e. the typical correlations between the spins in the magnet diverges. The diverging correlation length is indeed necessary for the renormalization procedure to work. The renormalization approach is similar to a mean field description, however self-similarity implies that the renormalization procedure is exact: we study not the original system at some length scale l_1 with a large number of degrees of freedom, but a rescaled one at length l_2 which gives us exactly the same physics.

---

[8] Note that here I only claim that this reduction is successful in so far as it allows quantitative predictions of certain macro-variables. I do not address the question as to whether the micro-model offers an encompassing explanation, as discussed, for example, in Batterman (2001).

This invariance that is based on a diverging correlation length is what distinguishes the Ising model from the Navier-Stokes case and makes it similar to the semi-classical laser theory insofar as we can reformulate the diverging correlation length as an extreme case of a separation of scales. A diverging scale can be understood as a limiting case of scale separation: one scale goes to infinity and the behavior of the whole system is determined by only this scale as the other degrees of freedom are, in the language of non-equilibrium phase transitions, slaved by the diverging scale.[9]

5.5 Perturbative Methods and Local Scale Separation

Discussions of various branches of physics have shown how scale separation is a universal feature of an explanatory successful micro-reduction. In this Section I briefly sketch how to generalize the concept of scale separation to the weaker requirement of *local* separation of scales. The universal behavior we encountered in the preceding section in critical theory, for example, is then lost. But as will be shown, non-standard perturbative accounts as well as some standard techniques in turbulence-modeling actually build on local scale separation.

As the introductory example of the movement of the Earth around the Sun showed, more traditional methodologies which may not be connected with reduction, such as perturbative accounts, may also be cast in the terms of scale separation. Such perturbation methods work when the perturbation is small or, in other words, the impacts of the higher perturbative expansions are smaller by orders of magnitude than the leading term.

In singular perturbation theory, as applied, for example, to so-called 'layer-type problems' within fluid dynamics, the condition that the perturbation is 'small' is violated within one or several layers at the boundary of the flow or in the interior of the spatio-temporal domain. The standard perturbative account that naively sets the perturbation to zero in the whole domain fails as this would change the very nature of the problem. The key to solving such problems—which is lucidly illustrated in Prandtl's (1905) boundary layer theory—is to represent the solution to the full problem as the sum of two expansions: loosely speaking, the solution in one domain is determined by the naive perturbative account, while the solution in the other domain is dominated by the highest derivative. I want to refer to this and the following examples as 'local scale separation' as it is still the separation of the relevant scales that allows for this approach; however, the scales do not separate on the entire domain.[10]

---

[9] This is the reason why, strictly speaking, the renormalization procedure works only at the critical point. The more remote the system is from the critical point, the more dominant become finite fluctuations of the correlation length that spoils the self-similarity.

[10] Thanks to Chris Pincock for pointing out this aspect.

Regular perturbation theory can also fail when the small perturbations sum up; this happens when the domain is unbounded. The techniques used here—multiple scale expansion and the method of averaging—again draw on local scale separation. For a technical account on the non-standard perturbative methods discussed here see Kevorkian and Cole (1996).

But local scale separation is not just the decisive feature that makes non-standard perturbative approaches work, it is also used more broadly in scientific modeling. An example invovles the 'large-eddy simulations', an approach to fluid dynamic turbulence that is very popular within the engineering sciences. Here the large-scale unsteady turbulent motion is represented by direct numerical integration of the Navier-Stokes equations that are believed to describe the fluid motion. Direct numerical simulation of all scales is computationally too demanding for most problems; thus within the large-eddy simulations the impact of the small-scale motion of the eddies is estimated via some heuristic model. This separate treatment of the large and the small scale motion is only possible when the so-called 'inertial range' (see Fig. 5.1) is large enough so that the motion of the smallest eddies inside the dissipation range, which are expected to have universal character in the sense that they are not influenced by the geometry of the flow, and the motion of the largest eddies inside the energy containing range can indeed be separated. Cast in terms of scale-separation, large-eddy simulations thus work when the typical scales of dissipative range are much smaller than those of the energy range (cf. Fig. 5.1; Pope 2000, p. 594f.).

5.6 Reduction, Emergence and Unification

The case studies in this paper exemplify how scale separation is the underlying feature of successful micro-reductionist approaches in different branches of physics, and how scale separation provides a unified view of various different methods used in diverse branches: the slaving principle, adiabatic elimination, standard and non-standard perturbative accounts, and critical theory. Success was thereby defined in Sect. 5.2 in epistemic terms: it amounts to deriving quantitative information of generalized state variables of the macro system from the micro-description. Also, in other areas of science where micro-reductionist models are employed and these models are formulated in mathematical terms, scale separation plays a crucial role.[11] In the remainder I want to address some of the implications the central role of scale separation has for the philosophical debates in condensed matter physics and complex systems theories, as well as the debates on reduction and modeling in the sciences more generally.

---

[11] Forest-fire models within ecology (Drossel and Schwabel 1992), or the Black-Scholes model in economics are such examples (Black and Scholes 1973).

Consider first the question of whether condensed matter physics, and particularly solid state physics, as well as complex systems analysis, occupy a special position within the physical sciences. It was shown that scale separation is the underlying feature not only when applying the slaving principle or adiabatic elimination but also in perturbative accounts or in QED. This clarifies why the often articulated idea that condensed matter physics or complexity theories play a special methodological role cannot be defended as a general claim (see Kuhlmann 2007; Hooker et al. 2011).

What results is a kind of scepticism regarding a reorganization of the sciences according the various methods used by each (e.g. Schmidt 2001). To some extent this may not be a fruitful line to pursue because for many methods the underlying idea, namely scale separation, is the same in different areas: seemingly different methodologies like adiabatic elimination or the slaving principle are just one way of expressing the fundamental fact that the system under consideration exhibits scale separation.

As mentioned before, well-established analytic tools like the slaving principle in Haken's theory of synergetics are seen by proponents of a reductionistic view as clear-cut examples for reduction, while anti-reductionists use the very same examples to support the contrary point of view. Identifying scale separation as the characteristic feature of an explanatory successful micro-reduction, adds more nuance to the picture: the analysis of the semi-classical laser theory was able to reveal how the slaving principle, adiabatic elimination, and others are good examples of methodological reduction. Moreover it showed that the application of the scaling principle turns an undisputed ontological micro-reduction into a successful explanatory one. Micro-reduction with scale separation thus provides not only the basis for a unified methodology used over various branches of the mathematical sciences, but also an important specific class of an epistemic or explanatory reduction. It also showed that questions of ontological reduction must be clearly separated from epistemic or explanatory aspects. Despite uncontested micro-reduction on an ontological level, complex systems without scale separation may provide examples of emergent epistemic features. Fluid dynamic turbulence provides a case at hand. I hence want to adopt a pluralistic stance concerning the question of epistemic priority of either the macro or micro level.[12]

This epistemic pluralism has important implications for contemporary computational sciences. It was the improvement of numerical methods and the sharp increase in computational power over the last decades that boosted the power of micro-reductionist approaches: computers made it possible to calculate micro-evolution equations that are more accessible than the equations that describe the maco behaviour. A field like solid-state physics owes large parts of its success to the employment of computers. The same is true for many other fields. For example, contemporary climate models try to resolve the various feedback mechanisms

---

[12] Compare Hüttemann (2004) for a critical discussion on the ontological priority of the micro level.

between the independent components of the overall climate system in order to predict the macro quantity 'global mean atmospheric temperature'. For models with such practical implications it becomes particularly clear why information about the stability of the system does not suffice, rather, what is needed are generalized state variables at the macro level.

Hence, the type of information many computer simulations aim for are quantitative predictions of generalized state variables. The use of these models, however, often does not distinguish between an epistemic or explanatory successful reduction and an ontological reduction. As my analysis has shown, even for cases where the ontological reduction is undisputed, as in condensed matter physics or climatology, there are serious doubts whether a micro-reductionist explanation is successful if the scales in the micro-model do not separate. This is the case for certain fluid dynamic problems and hence for climatology. The problem casts doubt on the often unquestioned success of numerically implemented micro-models.

This doubt also seems to be shared by scientists who have begun to question the usefulness of what appears to be the excessive use of computer simulations. In some fields where the reductionist paradigm was dominant one can see the first renunciations (e.g. Grossmann and Lohse 2000) of this approach. Just as it is not helpful to describe the motion of a pendulum by the individual motion of its 10^{23} atoms, micro-reductionist models that are numerically implemented are not a panacea for every scientific problem, despite the undisputed ontological reduction to the micro-description.

References

Batterman, R.W.: The Devil in the Details: Asymptotic Reasoning in Explanation, Reduction and Emergence. Oxford University Press, New York (2001)

Black, F., Scholes, M.: The pricing of options and corporate liabilities. J. Polit. Econ. **81**, 637–654 (1973)

Drossel, B., Schwabl, F.: Self-organized critical forest-fire model. Phys. Rev. Lett. **69**, pp. 1629–1632 (1992)

Falkovich, G., Sreenivasan, K.R.: Lessons from hydrodynamic turbulence. Phys.Today **59**(4), 43–49 (2006)

Frisch, U.: The Legacy of A. N. Kolmogorov. Cambridge University Press, Cambridge (1995)

Grossmann, S., Lohse, D.: Scaling in thermal convection: a unifying theory. J. Fluid Mech. **407**, 27–56 (2000)

Haken, H.: Synergetics. Introduction and Advanced Topics. Springer, Berlin, Heidelberg (2004)

Hooker, C.A., Gabbay, D., Thagard, P.,Woods, J.:Philosophy of Complex Systems, vol. 10. Handbook of the Philosophy of Science, Elsevier, Amsterdam (2011)

Hoyningen-Huene, P.: Epistemological reduction in biology: intuitions, explications and objections. in: Hoyningen-Huene, P., Wuketits, F.M. (eds.) Reductionism and Systems Theory in the Life Sciences, pp. 29–44. Kluwer, Dordrecht (1989)

Hoyningen-Huene, P.: On the way to a theory of antireductionist arguments. in: Beckermann, A., Flohr, H., Kim, J. (eds.) Emergence or Reduction? Essays on the Prospects of Nonreductive Physicalism, pp. 289–301. de Gruyter, Berlin (1992)

Hoyningen-Huene, P.: Reduktion und Emergenz. in: Bartels, A., Stöckler, M. (eds.) Wissenschaftstheorie, pp. 177–197. Mentis, Paderborn (2007)

Hütteman, A.: What's Wrong With Microphysicalism?. Routledge, London (2004)

Kuhlmann, M.: Theorien komplexer Systeme: Nicht-fundamental und doch unverzichtbar,. in: Bartels, A., Stöckler, M. (eds.) Wissenschaftstheorie, pp. 307–328, Mentis, Paderborn (2007)

Kevorkian, J., Cole, J.D.: Multiple Scale and Singular Perturbation Methods. Springer, New York (1996)

Kim, J.: Epiphenomenal and supervenient causation. Midwest Stud. Philos 9, 257–270 (1984); reprinted in: Kim J.: Supervenience and Mind, pp. 92–108. Cambridge University Press, Cambridge

Morgan, M., Morrison, M.C. (eds.): Models as Mediators: Perspectives on Natural and Social Science, Cambridge University Press, Cambridge (1999)

Nagel, E.: The Structure of Science. Routledge and Kegan Paul, London (1961)

Oseledets, V.: A multiplicative ergodic theorem. Characteristic Lyapunov exponents of dynamical systems. Trudy Moskov. Mat. Ob 19, 179 (1968)

Poincaré, J.H.: Leçons de mécanique céleste : professées à la Sorbonne, Tome I: Théorie générale des perturbations planétaires. http://gallica.bnf.fr/ark:/12148/bpt6k95010b.table (1893)

Pope, S.B.: Turbulent Flows. Cambridge University Press, Cambridge (2000)

Prandtl, L.: Über Flüssigkeitsbewegung bei sehr kleiner Reibung. In: Verhandlungen des dritten Internationalen Mathematiker-Kongresses, pp. 484–491. Heidelberg, Teubner, Leipzig (1905)

Schmidt, J.C.: Was umfasst heute Physik?—Aspekte einer nachmodernen Physik. Philosophia naturalis 38, 271–297 (2001)

Stöckler, M.: Reductionism and the new theories of self-organization. In: Schurz, G., Dorn, G. (eds.) Advances of Scientific Philosophy, pp. 233–254. Rodopi, Amsterdam (1991)

Wells, J.: Effective Theories in Physics. Springer, Dordrecht (2013)

Wimsatt, W.C.: Reductive explanation: A functional account. In: Cohen, R.S., Hooker, C.A., Michalos, A.C., van Evra, J.W. (eds.) PSA 1974, Proceedings of the 1974 Biennial Meeting, Philosophy of Science Association, pp. 671–710. Reidel, Dordrecht, Boston (1976)

Zaikin, N., Zhabotinsky, A.: Concentration wave propagation in two-dimensional liquid phase self oscillating system. Nature 225, 535–537 (1970)

Part II
Emergence

Chapter 6
Why Is More Different?

Margaret Morrison

6.1 Introduction

An emergent property or phenomenon is usually defined as one that arises out of lower level constituents but is neither reducible, explainable nor predictable from them. Emergence is sometimes associated with non-reductive physicalism, a view that advocates the physical nature of all concrete entities while acknowledging that some entities/properties that arise from this physical base cannot be reduced to it. The philosophical challenge is how to understand the relation between these various ontological and explanatory levels, especially since emergentists claim a distinct status for emergent phenomena/properties, distinguishing them from straightforward aggregates of constituents. To use Anderson's words, the whole is not only greater than but very *different* from the sum of the parts (1972, p. 395).

When dealing with emergence in physics, physicalism is not an issue. No one denies that emergent phenomena in condensed matter physics (e.g. superconductivity) are comprised of elementary particles or are physical in nature. Rather, the concern is whether some variant of reduction is really at work in contexts typically associated with emergence. An advocate of reduction could easily claim that because the macro level is composed of micro constituents there is no physical difference between different levels; instead what is lacking is an appropriate type of explanatory relation. Consequently, appeals to emergence simply indicate insufficient knowledge of the relevant explanatory connections between different theoretical levels, not a physical difference.

Moreover, the definition of emergence given above, which is the one commonly used in most discussions of emergence, is fully satisfied on purely epistemological grounds; further suggesting that emergence may simply point to a gap in our

M. Morrison (✉)
Department of Philosophy, Trinity College University of Toronto, Toronto,
ON, M5S 1H8, Canada
e-mail: mmorris@chass.utoronto.ca

© Springer-Verlag Berlin Heidelberg 2015
B. Falkenburg and M. Morrison (eds.), *Why More Is Different*,
The Frontiers Collection, DOI 10.1007/978-3-662-43911-1_6

knowledge. In keeping with this epistemic orientation we also find emergence described in terms of novelty. For example, Butterfield (2011) defines emergence in terms of "behaviour that is novel and robust relative to some comparison class". In other words, we should understand the properties or behaviour of a composite system as novel and robust compared to its components. Defining emergence in this way requires that we carefully distinguish between phenomena that are properly emergent and those that are simply aggregates. In the latter case we can reduce the composite to its constituents as in the case of a house that can be decomposed into the various materials used to build it.

However, what Anderson's characterization suggests is that emergence has a strongly ontological dimension and indeed examples of emergence in physics tend to support this way of thinking. A philosophical account of emergence that is ontologically based requires that we lay out, in an explicit way, how the micro and macro levels in emergent behaviour/phenomena are related. In other words, what causal role does the microphysics play in characterizing emergent phenomena and does this relation presuppose some implicit type of reductionism?

Perhaps the most important feature in characterizing the micro/macro relation in emergence is the notion of autonomy and the supposed independence of these two levels in explaining emergent behaviour. The relation between ontological and epistemic independence is especially important since the latter is a necessary but not a sufficient condition for emergence; the fact that we *need not* appeal to micro phenomena to explain macro processes is a common feature of physical explanation across many systems and levels. Instead, what is truly significant about emergent phenomena is that we supposedly *cannot* appeal to microstructures in explaining or predicting these phenomena even though they are constituted by them.

I begin by reviewing some arguments that address the issue of autonomy and ontological aspects of emergence (e.g. Howard 2007; Humphreys 1997a, b) and discuss why they fail to capture the features necessary for emergent phenomena in physics. From there I go on to discuss the relation between emergence and phase transitions and why we need an account of emergence at all. As an illustration of the micro/macro relation I focus on superconductivity and how it is possible to derive its characteristic features, those that define a superconductor (infinite conductivity, flux quantization and the Meissner effect), simply from the assumption of broken electromagnetic gauge invariance. I end with a brief discussion of the relation between physics and mathematics and its relevance for emergence. Emphasising the importance of emergence in physics is not to deny that reductionism has been successful in producing knowledge of physical systems. Rather, my claim is that as a global strategy it is not always capable of delivering the information necessary for understanding the relation between different levels and kinds of physical phenomena. As such, emergence becomes an important part of how we come to understand fundamental features of the physical world.

6.2 Autonomy and the Micro/Macro Relation: The Problem

In discussions of emergence in physics it is important to keep in mind that the problem is articulating the relationship between different levels of phenomena, not their ontological status. Emergent physical phenomena are typically thought to exhibit new causal "powers", meaning that new physics emerges at different energy or length scales. Philosophical debates about emergence have often appealed to non-reductive physicalism as a way of capturing the autonomy of emergent phenomena, with supervenience being the preferred way of describing the micro/macro relation. The appeal of supervenience is that it allows one to retain the beneficial features of reduction without embracing its difficulties, that is, without having to say, exactly, what the relation between x and y is, over and above the fact that the latter supervenes on the former.[1]

There are several accounts of supervenience but most involve a type of dependency relation where the lower-level properties of a system determine its higher level properties. The relation is often characterized in the following way: A supervenes upon another set B just in case no two things can differ with respect to A-properties without also differing with respect to their B-properties. In slogan form, "there cannot be an A-difference without a B-difference". Since we can assume, for the context of this discussion, that physicalism is unproblematic, an extended discussion of the pros and cons of supervenience needn't concern us at this point.[2] Instead I want to briefly look at some ontological accounts of emergence, each of which respects the "autonomy" of emergent phenomena by showing why the identity claims characteristic of reduction fail. The question that concerns me is whether these accounts of autonomy can successfully capture features associated with emergent phenomena in condensed matter physics, the field where emergent phenomena are perhaps most evident.

Humphreys (1997a, b) defines emergence in terms of a fusion relation operating between different levels of entities and properties.[3] He characterises the fusion relation (1997, p. 8) by defining a class of i-level properties and entities, P_m^i and x_r^i respectively, as the first level at which instances of $P_m^i\left(x_r^i\right)$ occur. The fusion operation

---

[1] Rueger's (2000) account of emergence involves a notion of supervenience defined in terms of stability or robustness. An emergent phenomenon/property is produced when a change in the subvenient base produces new behaviour that is both novel and irreducible. The causal powers that emergent phenomena have are simply those that "structural properties have in virtue of being configurations of their lower level constituents" (2000, p. 317). My difficulty with this view is that even if the emergent properties are novel and irreducible they are still the result of the system configured in a certain way. Consequently the causal powers of the whole are no different from those of the parts, making emergent properties similar to resultant properties.

[2] See Beckermann, Flohr and Kim (1992) for various discussions.

[3] If we identify emergent properties as resulting from the interaction of the constituents then it isn't immediately clear how to motivate the "more is different" claim characteristic of emergent phenomena.

[*] results in the following: if $P_m^i(x_r^i)(t_1)$, $P_n^i(x_s^i)(t_1)$ are i-level property instances, then $[P_m^i(x_r^i)(t_1) * P_n^i(x_s^i)(t_1)]$ is an i + 1-level property instance, the result of fusing $P_m^i(x_r^i)(t_1)$ and $P_n^i(x_s^i)(t_1)$. According to Humphreys it is the physical interactions represented by the fusion operation that lead to the transition from the i to $i + 1$ level that is responsible for emergent features. The fused $[P_m^i * P_n^i][(x_r^i) + (x_s^i)](t_1')$ is a unified whole in that its causal effects cannot be correctly represented in terms of the separate causal effects of $P_m^i(x_r^i)(t_1)$ and $P_n^i(x_s^i)(t_1)$. Moreover, within the fusion $[P_m^i * P_n^i][(x_r^i) + (x_s^i)](t_1')$ the original property instances $P_m^i(x_r^i)(t_1)$, $P_n^i(x_s^i)(t_1)$ no longer exist as separate entities and do not possess all their i-level causal powers available for use at the $(i + 1)$ level. In other words, these i-level property instances no longer have independent existence within the fusion; they simply go out of existence in producing the higher level emergent instances.

Here the subvenient base cannot be the reason why the emergent property is instantiated since the $i + 1$ level property instances do not supervene upon the i-level property instances. Humphreys (15) cites the example of quantum entanglement as a case of emergence resulting from the kind of fusion he describes. The composite system can be in a pure state when the component systems are not, and the state of one component cannot be completely specified without reference to the state of the other component. He sees the interactions that give rise to the entangled states as having the features required for fusion because the relational interactions between the constituents can no longer be separately individuated within the entangled pair.[4]

Silberstein and McGeever (1999, p. 187) claim that "QM provides the most conclusive evidence for ontological emergence" and their discussion of entanglement (189) appears to endorse the appropriateness of fusion for describing the whole-part relation in this context. In Howard (2007, p. 12) the paradigm case of emergence is also quantum entanglement and he claims that in areas of condensed matter physics where there is a reasonably successful theory (superconductivity and superfluidity) there is also a clear connection to microphysical entanglement. As an example he cites the role of Bose-Einstein condensates (BEC) in superfluidity and the way that Cooper pairs in superconductivity are, in effect, BECs.[5] Consequently the phenomena of condensed matter physics supervene on the most basic property of the micro-realm—entanglement (17). Howard states that "while condensed

---

[4] Humphreys also discusses examples of emergent phenomena that aren't of this sort, namely those that occur in ideal *macroscopic* systems containing an infinite number of particles (1997b, p. 342). His point is that the emergent properties cannot be possessed by individuals at the lower level because they occur only with infinite levels of constituents. Since these are exactly the sorts of examples I will have more to say about below.

[5] A Bose-Einstein condensate is a state of mater formed by bosons confined in an external potential and cooled to 0 kelvin or -273.15 °C. This causes a large fraction of the atoms to collapse into the lowest quantum state of the external potential.

matter physics does not obviously reduce to particle physics, phenomena...such as superfluidity and superconductivity do supervene on physical properties at the particle physics level and hence are not emergent with respect to particle physics..." (6). In other words, "the physical structure that [does] the explaining in condensed matter physics....is entanglement...the micro-world upon which condensed matter physics is said not to supervene" (22).

While these various claims about entanglement as an example of emergence are certainly plausible, the converse is less convincing; in other words, phenomena such as superconductivity, crystallization, magnetization, superfluidity, are neither explained by nor ontologically identified with quantum entanglement.[6] Nor can the latter account for the stability associated with these phenomena and the ability to make very accurate predictions about their behaviour. Humphreys is explicit that emergence does not require supervenience insofar as the fused properties cease to exist once the emergent phenomenon is present. But, how can this enable us to retain the ontological independence of the micro level in contexts like CMT and particle physics? Howard's solution is to understand these relations as supervenient, but this is of little help if we understand supervenience in the typical way, where the connection between the two levels requires a covariance relation to be maintained.[7] While Howard acknowledges that supervenience does not imply reduction, we shall see below that the kind of phenomena considered emergent in CMT, specifically universal behaviour, is not actually explained in terms of microphysical properties in the way he suggests, nor does it exemplify a supervenience relation. The characteristic behaviour(s) that identify phenomena as emergent (e.g. infinite conductivity) are neither explained nor identified with microphysical constituents.[8] Moreover, one of the hallmarks of emergent phenomena is that they are insensitive to their microphysical base which challenges the dependency relation present in supervenience.

Because emergent phenomena 'arise out' of their microphysical base we need some account of the ontological connection between the levels to fully explain the exact nature of the 'emergence' relation. In the case of ontological reduction there exists a type of *identity* that cannot be upheld in cases of emergence. Reductionism assumes, among other things, that because a particular macro phenomenon is a

---

[6] Although entanglement is undoubtedly operating here my use of the term 'identified' is meant to indicate that I don't subscribe to the view that emergent phenomena are *explained* via an ontological identification with entangled states, nor does the association with entanglement serve as an example of the supervenience relation where the basal property is associated with the higher level property.

[7] Howard cites Davidson's (1970) definition where supervenience is described as an ontic relationship between structures.construed as a set of entities. The higher level (B) entities supervene on the lower level (A) ones iff the former are wholly determined by the latter such that any change in (B) requires a corresponding change in (A).

[8] Infinite conductivity is one of the properties, along with flux quantization and the Meissner effect, that are exact regardless of the type of metal that comprises the superconductor.

collection of micro entities/properties the latter not only explains the behaviour of the former, giving us some insight as to why it behaves as it does, but it also constitutes it. Emergence shows us that the opposite is true! Initially this appears somewhat confusing because, for example, we typically understand the causal foundation of superconductivity in terms of Cooper pairing; so to claim that there is no reduction to or identification with this microphysical base requires a clarification of the exact nature of these ontological relations.

In what follows I show how the nature of universality as well as the role played by it clarifies both how and why emergent phenomena are independent of any specific configuration of their microphysical base. An important advantage of this strategy is that the microphysical entities and properties remain intact and autonomous, unlike Humphreys' fusion relation or accounts that appeal to quantum entanglement. As we shall see below, this relative independence from the underlying microphysics is crucial for understanding the difference between emergent and resultant properties and for highlighting the similarities and differences between emergence and multiple realizability.

6.3 Emergence and Reduction

In physics it has been common to think of explanation in reductivist terms, involving the elementary constituents of matter and the laws that govern them. Indeed this is the motivation behind a good deal of contemporary physics and is a strategy that has not been without success, as in the case of Maxwell's electrodynamics and Newtonian mechanics. Although the limits and difficulties associated with various forms of reductionism (ontological and inter-theoretic) have been well documented, it is still thought of as the ultimate form of explanation, as something to aspire to despite the difficulties attaining it.[9]

When evaluating the merits of reductionist explanation it is also important to inquire about its limits and how far this kind of explanation extends, specifically, what actually counts as "reduction" and at what point does the addition of free parameters undermine reductionist claims? The non-relativistic Schrodinger equation presents a nice illustration of the kind of reduction we typically associate with explanation in physics.

$$i\hbar \frac{\partial}{\partial t} |\Psi> = \mathscr{H} |\Psi>$$ (6.1)

---

[9] This is especially true in the philosophy of science literature. Sklar has written extensively on the problems of reduction and the relation between thermodynamics and statistical mechanics. See his (1999) for a pointed discussion of these issues.

It describes, in fairly accurate terms, a large number of physical phenomena and can be completely specified by a few quantities such as the charges and masses of the electron and atomic nuclei, as well as Planck's constant. It can be solved accurately for small numbers of particles (isolated atoms and small molecules) and agrees in minute detail with experiment. However, as Laughlin and Pines (2000) point out, when the number of particles exceeds around ten this is no longer the case. It is possible to perform approximate calculations for larger systems which yield knowledge of atomic sizes and the elastic properties of solid matter, etc. but the use of approximation techniques means that these are no longer deductions from first principles or fundamental theory—instead they require experimental input and specific, local details. What this indicates is a breakdown of the reductionist ideal of deriving explanations of a large number of phenomena from a few simple equations or laws.

But does this *really* undermine reduction as an explanatory strategy? The answer depends, in part, on how many free parameters one is willing to accept into the explanation; in other words, at what point does it no longer make sense to call an explanation reductive when the explanatory information comes via the free parameters rather than fundamental features of theories/laws. Of course one might also argue that calling a phenomenon "emergent" is simply a stop gap measure indicating we haven't yet hit on the right theoretical principles. The difficulty with this type of response is that it offers only a promissory note and fails to help us understand the phenomena/system under investigation. Put slightly differently: Our lack of understanding results, in the first instance, from a failure in the reductive strategy; hence the need for an alternative framework. Whether we might someday be able to perform the right sort of derivations or calculations from first principles is irrelevant for evaluating the merits of reduction in the cases where it currently fails to provide the relevant information.

But, when it comes to articulating the important features of emergence we need to move beyond the failure of reduction or limiting inter-theoretic relations since this too can be indicative of an epistemic problem. Moreover, if emergence simply means that a phenomenon at one level, characterized by a particular theory, fails to be fully explainable by the theory at the next lower level then it becomes much too pervasive. Instead the focus should be on what is ontologically distinct about emergent phenomena such that they are immune from the contingencies of reduction.

Laughlin and Pines (2000) point out that the parameters $e, \hbar$, and m appearing in the Hamiltonian for the Schrodinger equation

$$\mathcal{H} = -\sum_{j}^{N_e} \frac{\hbar^2}{2m} \nabla_j^2 - \sum_{\alpha}^{N_i} \frac{\hbar^2}{2M_\alpha} \nabla_\alpha^2 - \sum_{j}^{N_e} \sum_{\alpha}^{N_e} \frac{Z_\alpha e^2}{|\vec{r}_j - \vec{R}_\alpha|} + \sum_{j<k}^{N_e} \frac{e^2}{|\vec{r}_j - \vec{r}_k|} + \sum_{j<\beta}^{N_j} \frac{Z_\alpha Z_\beta e^2}{|\vec{R}_\alpha - \vec{r}_\beta|}.$$

$$(6.2)$$

can be accurately measured in laboratory experiments involving large numbers of particles but can't be derived or predicted by direct calculation.[10] For example, electrical measurements performed on superconducting rings can determine, to a very high level of accuracy, the quantity of the quantum of magnetic flux $hc/2e$ and four point conductance measurements on semiconductors in the quantum Hall regime accurately determine the quantity e^2/h. Because it is impossible to derive these exact results using either first principles, or approximation techniques, the natural question that arises is what explains the stable behaviour in these cases?

Although no reductive explanation is possible the examples indicate, in a more pointed way, the need for 'emergence' in order to account for the stability. Laughlin and Pines claim that these type of experiments work because "there are higher organizing principles in nature that make them work" (2000, p. 28). Examples of such principles are continuous symmetry breaking which renders exact the Josephson quantum and localization which is responsible for the quantum Hall effect.[11] They claim that both effects are "transcendent" in that neither can be deduced from the microphysics and would continue to be true even if the theory of everything were changed. These are classified as emergent phenomena or 'protectorates'.

When Laughlin speaks of organizational principles he seems to have in mind the kind of order that is produced as a result of some type of collective action that is essentially independent of the details of the underlying microphysics. For example, he mentions principles governing atomic spectra that can be understood without any reference to the quark structure of nucleons and the laws of hydrodynamics which would be roughly the same regardless of variations in detailed intermolecular interactions. However, in both of these cases we need to differentiate explanatory from ontological claims since emergence isn't simply about different organizational principles being important at different scales or laws not requiring specific micro-details. More is required.

In Laughlin's and Pines' discussion of continuous symmetry breaking they don't elaborate on the notion of 'transcendence' or the status of organizing principles, but in the latter case independence from specific theoretical content is going to be necessary if emergent phenomena are to be properly autonomous from the microphysical domain. While many physical theories/phenomena incorporate or involve various types of symmetry breaking, the notion itself is not linked to any *specific* theoretical framework. Rather, it functions as a structural constraint on many different kinds of systems in both high energy physics as well as condensed matter physics.

---

[10] The symbols $Z\alpha$ and $M\alpha$ are the atomic number and mass of the αth nucleus, $R\alpha$ is the location of this nucleus, e and m are the electron charge and mass, r_j is the location of the jth electron, and h is Planck's constant.

[11] Localization involves the absence of diffusion of waves in a random medium caused by a high concentration of defects or disorder in crystals or solids. In the case of electric properties in disordered solids we get electron localization which turns good conductors into insulators.

I refer to symmetry breaking as a "structural/dynamical feature of physical systems" because of the way order and structure emerge as a result of the phase transitions associated with symmetry breaking. In fluid dynamics the emergence of new order and structure occurs when a dynamical system is driven further and further away from thermal equilibrium. By increasing control parameters like temperature and fluid velocity old equilibria become unstable at critical points, break down, and new branches of local equilibria with new order emerge. Spontaneous symmetry breaking (SSB) is manifest in, among other things, the acquisition of rigidity and the existence of low energy excitations in condensed matter physics; superconductivity incorporates symmetry breaking via Cooper pairing as a basic feature in the BCS theory. Particle masses in high energy physics are also thought to be generated by SSB. In each of these cases we have dynamical processes that produce specific effects. Because these processes involve a connection with microphysics, the challenge for the emergentist is to explain how and why we should think of symmetry breaking as distinct from the type of fundamental physics associated with reduction. We'll see why this is the case below.

Before discussing that point it is important to mention that the status of SSB in the case of local gauge symmetries (of the kind relevant for superconductivity) is not entirely clear. Elitzur's theorem (1975) states that local gauge symmetries cannot be spontaneously broken. Although the theorem was proved for Abelian gauge fields on a lattice it is suggested that it doesn't rule out spontaneously broken global symmetries within a theory that has a local gauge symmetry, as in the case of the Higgs mechanism. There is a good deal of controversy regarding the interpretation of SSB as a "physical" phenomenon with the main arguments enumerated and addressed by Friedrich (2013) who also argues against the realistic interpretation. Although I certainly cannot provide a proper discussion of the issue here, let me mention a few points worth keeping in mind regarding the role SSB plays in the theoretical context of phase transitions.

First, it is important to note that Elitzur's theorem is specific to the lattice because on the lattice it isn't necessary to fix a gauge. Moreover, many claim that the lattice description is the appropriate one because it eliminates any reliance on perturbation theory. While there are conflicting pictures presented by the continuum and lattice formulations (see Frolich et al. 1981) one further point is worth emphasising from the "realist" perspective. It is certainly possible to carry out perturbative calculations with a Lagrangian having a local symmetry in which scalar fields that are not invariant under the symmetry have non vanishing vacuum expectation values (VEVs). This, it would seem, deserves to be called a spontaneously broken local symmetry. Perhaps the difficulty and confusion surrounding this issue arises as a result of perturbation theory; nevertheless, let me assume for the sake of argument that SSB does in fact occur in phase transitions. What are the interesting implications for emergence?

When symmetries are spontaneously broken the result is the occurrence of ordered states of the sort Laughlin refers to. For example, magnetisation results from broken spin rotation symmetry and massive particles break a phase rotation symmetry. These symmetries impose structural constraints on the physical world in that they give rise to and explain certain forms of dynamical interactions. As we shall see below these constraints are *general* structural features of physical systems that can apply in a variety of theoretical contexts. By contrast, fundamental theory is concerned with details, expressed via laws and models, of *specific* physical systems and how they behave. It is these general features rather than specific details of micro-processes that prove important for emergence. In order to clarify the ontological relations among emergence, symmetry breaking and microphysics let me turn to the example of superconductivity which nicely illustrates these features.

6.4 Phase Transitions, Universality and the Need for Emergence

As we saw above one of the organizing principles Laughlin and Pines mention is continuous symmetry breaking. While many physical theories/phenomena incorporate or involve various types of symmetry breaking the notion itself is not linked to any *specific* theoretical framework. Rather, it functions as a structural constraint on many different kinds of systems in both high energy physics as well as condensed matter physics. For example, the electroweak theory postulates symmetry breaking via the Higgs mechanism which allegedly explains bosonic masses; superconductivity also incorporates symmetry breaking via Cooper pairing as a basic feature in the BCS theory. Because these processes appear to involve a connection with microphysics, the challenge is to explain how and why we should think of symmetry breaking as an organizing principle and not part of "fundamental" theory.

Other types of organizing principles like kinship and valency function as either a principle for organizing individuals into groups or in the latter case as a measure of the number of chemical bonds formed by the atoms of a given element. Valency, understood as an organizing principle, has evolved into a variety of approaches for describing the chemical bond such as valence bond theory and molecular orbitals, as well as methods of quantum chemistry. In that sense it provides a foundational framework within which different methodological approaches can be unified and also functions as a kind of heuristic principle in the elementary study of covalent bonds. But, when Laughlin speaks of organizational principles he has in mind the kind of order that is produced as a result of some type of collective action that is essentially independent of the details of the underlying microphysics. For example, he mentions principles governing atomic spectra that can be understood without any reference to the quark structure of nucleons and the laws of hydrodynamics which would be roughly the same regardless of variations in detailed intermolecular interactions.

While this seems like a claim about different organizational principles being important at different scales, emergence isn't captured simply by an appeal to different levels nor do physical explanations always require an appeal to "fundamental" theories. True independence from fundamental theory, as characterised by emergence, requires that we locate the relevant explanatory details in more general features capable of explaining how emergent phenomena arise. Crystals arise from the breaking of translation symmetry, magnetisation from broken spin rotation symmetry and massive particles break a phase rotation symmetry. These symmetries impose structural constraints on dynamical features of the physical world described by our theories. To that extent they do more than simply organize phenomena into certain types, they function as meta-laws via their role in explaining certain forms of dynamical interactions. To the extent that symmetry breaking explains certain features/behaviour of physical phenomena we can distinguish it from the role that fundamental theory plays in explanation, explanations whose focus is microphysical phenomena and the laws that govern them. Maintaining this distinction is crucial for upholding the autonomy of emergent phenomena.

On a very basic level we can think of symmetry constraints as providing us with general structural principles that apply in a variety of theoretical contexts; fundamental theory, on the other hand, is concerned with more specific types of physical systems and the details of how those systems behave. Those details take the form of theoretical laws or models that describe and explain the behaviour of particular types of phenomena. For example, the Schrodinger equation and the Pauli exclusion principle are part of the theoretical framework of quantum mechanics, as are models like the finite potential well. By contrast symmetry principles like those mentioned above are associated with a wide variety of physical theories and laws, both quantum and classical and operate at a meta-theoretical level furnishing the very general features that systems possess. It is these general features rather than specific details of micro-processes that prove important for emergence. In order to clarify the sense in which this ordering could be thought of as 'transcendent' let me turn to the example of superconductivity to illustrate the relation between emergence, symmetry breaking and microphysics.[12]

Many of the physical properties of superconductors such as heat capacity and critical temperature (where superconducting properties are no longer present) vary depending on the type of metal. However, there is a class of properties that are independent of the specific material and are exact for all superconductors, properties such as infinite conductivity (very low electrical resistance and currents that can circulate for years without perceptible decay), flux quantization and the Meissner effect.[13] These can be predicted with extraordinary accuracy; but in deriving them and other phenomena associated with superconductors one typically uses models that are just reasonably good approximations. There are macroscopic models like Ginzburg-Landau where cooperative states of electrons are represented

---

[12] This is necessary especially as an answer to Howard (2007).

[13] The former is a quantum phenomenon in which the magnetic field is quantized in the unit of $h/2e$ while the latter simply refers to the explusion of a magnetic field from a superconductor.

using a complex scalar field and the microscopic model(s) of the Bardeen-Cooper-Schrieffer (BCS) theory where electrons appear explicitly and are assumed to interact only by single phonon exchange. The latter is the widely accepted account that explains the superconducting phase of a metal as involving many pairs of electrons (Cooper pairs) bound together at low temperatures. This pairing in a superconductor is a collective phenomenon analogous to magnetization in a magnet and, as with magnetism, involves symmetry breaking and a phase transition. The essence of the BCS theory is the appearance of a pair field which is the order parameter of the superconducting state, just as magnetization is the order parameter of the ferromagnet.[14] Exactly how this pairing occurs is the subject of different model explanations, one of which was provided by BCS themselves in their original paper (1957).[15]

It is tempting to see the story about Cooper pairing as a reductive, micro-causal explanation insofar as the electron pairs seem to be the defining characteristic of superconductivity. However, the story is more complicated than might first appear. Recall the discussion above of the Josephson effect and the problem of deriving exact results from approximations. The same situation arises with superconductivity where the properties (infinite conductivity, flux quantization etc.) are exact and the same for all superconductors. Since they are exact results they must follow from general principles rather than simply derived using approximations. So, while highly precise predictions about superconductors follow from the models they do so because the models embody a symmetry principle—the spontaneous breakdown of electromagnetic gauge invariance (Weinberg 1986, 1996). One needs detailed models like BCS to explain the specifics of *how* the symmetry breaking (SSB) occurs, at what temperature superconductivity is produced, and as a basis for approximate quantitative calculations, but not to derive the most important *exact* consequences of this breakdown—infinite conductivity, flux quantization and the Meissner effect—properties that define superconductors.[16]

This fact is crucial for our account of emergence because it shows that the microphysical details about *how* Cooper pairing takes place are not important in deriving and explaining fundamental features of superconductivity. Put differently, it isn't that instances of superconductivity in metals don't involve micro-processes, rather the characteristics that define the superconducting state are not explained or predicted from those processes and are independent of them in the sense that changes to the microphysical base would not affect the emergence of (universal) superconducting properties. Although the breakdown of gauge invariance involves the formation of Cooper pairs—a dynamical process—the micro story figures simply as the foundation from which superconductivity emerges.

---

[14] The order parameter is a variable that describes the state of the system when a symmetry is broken; its mean value is zero in the symmetric state and non-zero in the non-symmetric state.

[15] For more on the topic of superconductivity, theories and models see Morrison (2007, 2008).

[16] See Weinberg (1986, 1996) for details.

The key to understanding this relationship involves the connection between phase transitions and symmetries. Symmetry breaking is reflected in the behaviour of an order parameter that describes both the nature and magnitude of a broken symmetry. In the ferromagnetic state the order parameter is represented by the vector describing the orientation and size of the material's magnetization and the resulting field. In the superconducting case the order parameter is the amplitude $\langle \varphi \rangle$ of the macroscopic ground state wave function of the Cooper pairs. The electromagnetic properties in a superconductor are dominated by Cooper pairs whereas electrons in a metal normally behave as free particles that are repelled from other electrons due to negative charge. Because Cooper pairs only appear at T_c (their presence indicates that the system has undergone a phase transition) they give rise to the order parameter which implies that the Cooper pairs must form a single wave function. In general the order parameter can be thought of as an extra variable required to specify the macroscopic state of a system after the occurrence of a phase transition. In non-superconducting metals gauge invariance ensures that $\langle \varphi \rangle = 0$. It should be noted here that an order parameter can have a well defined phase in addition to an amplitude and it is the phase that governs the macroscopic properties of superconductors and superfluids.

Given this picture we now need to disentangle the relation between the order parameter and the emergent nature of superconductivity. Recall that the broken symmetry associated with the order parameter in superconductivity is electromagnetic gauge invariance. The electromagnetic properties are dominated by Cooper pairs with each pair j having a wave function

$$\psi_c^j(r) = V^{-1/2} a_j(r) \exp i\phi_j(r) \qquad (6.3)$$

where $a_j(r)$ and $\phi_j(r)$ represent the amplitudes and phases respectively. The mean separation at which pair correlation becomes effective is between 100 and 1,000 nm and is referred to as the coherence length, ξ, which is large compared with the mean separation between conduction electrons in a metal. In between one pair there may be up to 10^7 other electrons which are themselves bound as pairs. The coherence volume ξ^3 contains a large number of indistinguishable Cooper pairs so one must define a density of wave functions averaged over the volume. The average will only be non-zero if the phases $\phi_j(r)$ are close together; i.e. the neighbouring Cooper pairs are coherent. In the case of the groundstate wavefunction density

$$\psi(r) = 1/\xi^3 \sum_{j \in \xi^3} \psi_j(r_j) \propto \sqrt{n_s} \exp i\phi(r) \qquad (6.4)$$

we can identify $|\psi(r)|^2$ with the density of Cooper pairs at point r and then define creation and annihilation operators for particles at r. In a normal conducting metal the expectation value for these operators takes value zero but in superconductors the operator $\psi(r)$ acquires a non-zero expectation $\langle \psi(r) \rangle$. So, at zero temperature

$$\left\langle \psi^\dagger(r)\psi(r) \right\rangle = \left\langle \psi^\dagger(r) \right\rangle \left\langle \psi(r) \right\rangle. \tag{6.5}$$

The order parameter is then defined as the expectation value of operator $\psi(r)$.

Above I claimed that one can derive the exact properties of superconductors from the assumption of broken electromagnetic gauge invariance. To show that this is, in fact, the symmetry that is broken we consider the following: In a superconductor it is generally possible to choose the gauge $\Lambda(r)$ of the vector potential which determines the phase of the wave function of each particle, i.e.

$$\psi'(r) = \psi(r)\exp(2\pi i\Lambda(r)/\Phi_0),$$
$$A'(r) = A(r) + \nabla\Lambda(r) \tag{6.6}$$

If the particles are independent it is possible in principle to choose a different gauge to describe the motion of each particle. However, phase coherence between the various Cooper pairs requires that all the particles have the same gauge. Consequently, the symmetry broken by the order parameter is local gauge invariance. The same choice of vector potential must be made for all of the particles. The system thus selects a particular phase in the same way a magnet selects a particular direction below the Curie temperature. Choosing a particular phase for the order parameter amounts to choosing a particular gauge for the vector potential A—hence the physical significance of the electromagnetic gauge in this context.

We can now go on to show how to derive the exact (emergent) properties of superconductors from the assumption of broken electromagnetic gauge invariance. To demonstrate this we consider how the consequences of broken gauge invariance for superconductors can be derived from a formalism that deals solely with the general properties of the Goldstone mode which is a long-wavelength fluctuation of the corresponding order parameter.[17] The general framework is set up in the following way: The electromagnetic gauge group $U(1)$ is the group of multiplication of fields $\psi(x)$ of charge q with the phases $\psi(x) \rightarrow \exp(i\Lambda q/\hbar)\,\psi(x)$. Because the q are integer multiples of $-e$ the phases Λ and $\Lambda + 2\pi\hbar/e$ are taken to be identical. $U(1)$ is spontaneously broken to Z_2 the subgroup consisting of $U(1)$ transformations with $\Lambda = 0$ and $\Lambda = \pi\hbar/e$. According to the general understanding of SSB the system described by a Langranian with symmetry group G, when in a phase where G is broken to a subgroup H, will possess a set of Nambu-Goldstone excitations described by fields that transform under the symmetry group G like the coordinates of the coset space G/H. In this case there will be a single excitation described by a field $\varphi(x)$ that transforms under $G = U(1)$ like the phase Λ. The $U(1)$ group has the multiplication rule $g(\Lambda_1)g(\Lambda_2) = g(\Lambda_1 + \Lambda_2)$ so under a gauge transformation with parameter Λ, the field $\varphi(x)$ will undergo the transformation $\varphi(x) \rightarrow \varphi(x) + \Lambda$. Because $\varphi(x)$ parameterizes $U(1)/Z_2$ rather than $U(1)$, $\varphi(x)$ and $\varphi(x) + \pi\hbar/e$ are

---

[17] My discussion follows Weinberg (1986).

regarded as equivalent field values. The characteristic property of a system with broken symmetry is that the quantity $\varphi(x)$ behaves like a propagating field.

When one turns on the interaction of the superconductor with the electromagnetic fields **B** and **E** their interaction is governed by the principle of local gauge invariance where the Nambu-Goldstone field $\varphi(x)$ transforms under $U(1)$ with a space-dependent phase $\varphi(x) \rightarrow \varphi(x) + \Lambda(x)$. The potentials transform as usual and all the other field operators are gauge invariant. The Lagrangian for the superconductor plus electromagnetic field is:

$$L = \frac{1}{2} \int d^3x \left(E^2 - B^2 \right) + L_m[\nabla\varphi - A, \dot{\varphi} + A^0, \tilde{\psi}] \tag{6.7}$$

where the matter Lagrangian is an unknown function of the gauge invariant combinations of $\partial_\mu \varphi$ and A_μ as well as the unspecified gauge-invariants $\tilde{\psi}$ representing the other excitations of the system. From L_m one obtains the electric current and charge density as variational derivatives

$$J(x) = \delta L_m / \delta A(x) \tag{6.8}$$

$$\varepsilon(x) = -\delta L_m / \delta A^0(x) = -\delta L_m / \delta \dot{\varphi}(x). \tag{6.9}$$

Because $\varphi(x)$ is the only non gauge-invariant matter field we can use just the Lagrangian equations of motion for $\varphi(x)$ to derive the equation for charge conservation. The structure of the functional matter Lagrangian need not be specified, instead one need only assume that in the absence of external electromagnetic fields the superconductor has a stable equilibrium configuration with vanishing fields

$$\nabla\varphi - A = \dot{\varphi} + A^0 = 0. \tag{6.10}$$

The assumption that electromagnetic gauge invariance is spontaneously broken is equivalent to the claim that the coefficients of the terms in L_m of second order in $\nabla_\phi - A$ and $\dot{\varphi} + A^0$ have non-vanishing expectation values which makes φ behave like an ordinary physical excitation. As we shall see in deriving the consequences of these assumptions, the important point is that $\varphi(x)$ is not understood as the phase of a complex wave function used in an "approximate" model/treatment of electron pairing, but rather, a Nambu-Goldstone field that accompanies the breakdown of SSB. Put differently, we don't need a microscopic story about electron pairing and the approximations that go with it to derive the exact consequences that define a superconductor. Planck's constant $\hbar$ simply does not appear in the differential equations governing φ.

From this framework one can derive fundamental properties of superconductors like the Meissner effect, flux quantization and infinite conductivity. For example, in the case of flux quantization we have a current flowing through a superconducting loop in thick closed rings that is not affected by ordinary electrical resistance. It cannot decay smoothly but only in jumps. However, when dealing with infinite

conductivity one needs to take account of time-dependent effects. We saw above
(6.9) that charge density is given by $-\varepsilon(x) = \delta L_m/\delta\varphi(x)$ where $-\varepsilon(x)$ is the
dynamical variable canonically conjugate to $\varphi(x)$. In the Hamiltonian formalism
H_m is a functional of $\varphi(x)$ and $\varepsilon(x)$ with the time dependence of φ given by

$$\varphi(x) = \delta H_m/\delta(-\varepsilon(x)). \tag{6.11}$$

The voltage at any point is defined as the change in the energy density per change in
the charge density at that point

$$V(x) \equiv \delta H_m/\delta\varepsilon(x). \tag{6.12}$$

Consequently the time-dependence of the Nambu-Goldstone field at any point is
given by the voltage $\dot{\phi}(x) = -V(x)$. From this it follows that a piece of superconducting wire that carries a steady current with time independent fields must have
zero voltage difference between its ends, which is just what is meant by infinite
conductivity. Without this zero voltage the gradient $\nabla\varphi(x)$ would have to be time
dependent leading to time dependent currents or fields.

A crucial part of the story, which is significant for emergence, is the relation
between infinite conductivity and the presence of an energy gap in the spectrum of
the Cooper pairs. Typically it is the presence of an energy gap that distinguishes
superconductivity from ordinary conductivity by separating the Fermi sea of paired
electrons from their excited unpaired states. The process is thought to be due,
essentially, to quantum mechanics and it implies that there is a minimum amount of
energy ΔE required for the electrons to be excited. As temperature increases to Tc,
Δ goes to 0. Although some accounts of superconductors relate infinite conductivity
directly to the existence of the gap, the treatment above shows that infinite conductivity depends only on the spontaneous breakdown of electromagnetic gauge
invariance and would occur regardless of whether the particles producing the
pairing were fermions instead of bosons. This is further evidenced by the fact that
there are known examples of superconductors without gaps.

The advantages of thinking about emergence in this way is that it encompasses
and clarifies both the ontological and epistemological aspects. Although superconductors are constituted by their microscopic properties, their defining features
(infinite conductivity, flux quantization, the Meissner effect) are immune to changes
in those properties (e.g. replacing fermions with bosons). This is the sense in which
we can refer to the properties of a superconductor as 'model independent' and not
causally linked to a specific microphysical account. In other words, symmetry
breaking (here the breakdown of electromagnetic gauge invariance) provides the
explanation of emergent phenomena but the specific microphysical details of *how*
the symmetry is broken are not part of the account. In that sense the emergent
phenomenon is not reducible to its microphysical constituents yet both retain full
physical status. This also allows us to see why supervenience, understood in terms
of a dependency relation, is inapplicable in explaining the part-whole aspects of
emergent phenomena—there is no determining linkage between the micro and

macro levels. But this is exactly as it should be. What makes an emergent phenomenon emergent is that it satisfies certain conditions, one of which is that it can't be captured using a supervenience relation.

Although we can explain emergent phenomena in terms of the symmetry breaking associated with phase transitions, the physics inherent in this explanation is not entirely unproblematic (Bangu 2009; Callender 2001; Earman 2004; Menon and Callender 2013). A well known fact about phase transitions is that even though they take place in finite systems they can only be accounted for by invoking the thermodynamic limit $N \rightarrow \infty$. The link between assumptions about infinite systems and the physics of symmetry breaking/phase transitions is provided by renormalization group (RG) methods which function as a framework for explaining *how* certain types of phenomena associated with phase transitions arise, as well as the similarity in behaviour of very different phenomena at critical point (universality) (Wilson 1983). RG provides the interconnection between mathematics and physics; fleshing out those details will further exemplify the ontological independence of the microphysics in accounting for emergent phenomena.

6.5 Renormalization Group Methods: Between Physics and Mathematics

Part of the importance of the RG is that it shows not just that we can focus on the energies or levels we are interested in, leaving out the rest, as we sometimes do in idealization and model building; it also illustrates and explains the ontological and epistemic independence between different energy levels—the defining features of emergent phenomena. One of the hallmarks of a phase transition is that it exhibits the effects of a singularity over the entire spatial extent of the system. Theory tells us that this happens only in infinite systems (particles, volume or sometimes strong interactions) so phase transitions produce a variation over a vast range of length/ energy scales. As a mathematical technique RG allows one to investigate the changes to a physical system as one views it at different distance scales. This is related to a scale invariance symmetry which enables us to see how and why the system appears the same at all scales (self-similarity). As we saw above phase changes of matter are often accompanied by discontinuities such as magnetization in a ferromagnet. At critical point the discontinuity vanishes so for temperatures above Tc the magnetization is 0. We also saw that the non-zero value of the order parameter is typically associated with this symmetry breaking, so the symmetry of the phase transition is reflected in the order parameter (a vector representing rotational symmetry in the magnetic case and a complex number representing the Cooper pair wavefunction in superconductivity).

In RG calculations the changes in length scale result from the multiplication of several small steps to produce a large change in length scale l. The physical phenomena that reflect this symmetry or scale transformation are expressed in terms of observed quantities—mathematical representations of the symmetry operation. For example, quantities that obey rotational symmetry are described by vectors, scalars etc. and in the case of scale transformations power laws reflect the symmetries in the multiplication operations. The physical quantities behave as powers l^x where x can be rational, irrational, positive etc. Behaviour near critical point is described using power laws where some critical property is written as a power of a quantity that might become very large or small. The behaviour of the order parameter, the correlation length and correlation function are all associated with power laws where the "power" refers to the critical exponent or index of the system. Diverse systems with the same critical exponents (exhibiting the same scaling behaviour as they approach critical point) can be shown via RG to share the same dynamical behaviour and hence belong to the same universality class.

The correlation function $\Gamma(r)$ measures how the value of the order parameter at one point is correlated to its value at some other point. If Γ decreases very fast with distance, then far away points are relatively uncorrelated and the system is dominated by its microscopic structure and short-ranged forces. A slow decrease of Γ implies that faraway points have a large degree of correlation or influence on each other and the system thus becomes organised at a macroscopic level. Usually, near the critical point $(T \to T_c)$, the correlation function can be written in the form

$$\Gamma(r) \to r^{-p} \exp(-r/\xi) \tag{6.13}$$

where ξ is the correlation length. This is a measure of the range over which fluctuations in one region of space are correlated with or influence those in another region. Two points separated by a distance larger than the correlation length will each have fluctuations that are relatively independent. Experimentally, the correlation length is found to diverge at the critical point which means that distant points become correlated and long-wavelength fluctuations dominate. The system 'loses memory' of its microscopic structure and begins to display new long-range macroscopic correlations.

The iterative procedure associated with RG results in the system's Hamiltonian becoming more and more insensitive to what happens on smaller length scales. As the length scale changes, so do the values of the different parameters describing the system. Each transformation increases the size of the length scale so that the transformation eventually extends to information about the parts of the system that are infinitely far away. Hence, the infinite spatial extent of the system becomes part of the calculation and this behaviour at the far reaches of the system determines the thermodynamic singularities included in the calculation. The change in the parameters is implemented by a beta function

$$\{\tilde{J}_k\} = \beta(\{J_k\}) \tag{6.14}$$

which induces what is known as an RG flow on the J-space. The values of J under the flow are called running coupling constants. The phase transition is identified as the place where the RG transformations bring the couplings to a fixed point with further iterations producing no changes in either the couplings or the correlation length. The fixed points give the possible macroscopic states of the system at a large scale. So, although the correlation length diverges at critical point, using the RG equations reduces the degrees of freedom which, in effect, reduces the correlation length.

The important point that distinguishes RG from previous renormalization methods is that the number and type of relevant parameters is determined by the *outcome* of the renormalization calculation.[18] After a sufficient number of successive renormalizations all the irrelevant combinations have effectively disappeared leaving a unique fixed point independent of the value of all of the irrelevant couplings. Assuming that a fixed point is reached one can find the value that defines the critical temperature and the series expansions near the critical point provide the values of the critical indices.[19] The fixed point is identified with the critical point of a phase transition and its properties determine the critical exponents with the same fixed point interactions describing a number of different types of systems. In that sense RG methods provide us with physical information concerning how and why different systems exhibit the same behaviour near critical point (universality).

The basis of the idea of universality is that the fixed points are a property of *transformations* that are not particularly sensitive to the original Hamiltonian. What the fixed points do is determine the kinds of cooperative behaviour that are possible, with each type defining a universality class. The important issue here isn't just the elimination of irrelevant degrees of freedom, rather it is the *existence or emergence of cooperative behaviour* as defined by the fixed points. The coincidence of the critical indices in very different phenomena was inexplicable prior to RG methods. Part of the success of RG was showing that the differences were related to irrelevant observables—those that are "forgotten" as the scaling process is iterated. Another significant feature of RG is that it showed how, in the long wave-length/large space-scale limit, that the scaling process in fact leads to a fixed point when the system is at a critical point, with very different microscopic structures giving rise to the same long-range behaviour.

What this means for our purposes is that RG equations illustrate that phenomena at critical point have an underlying order. Indeed what makes the behaviour of critical point phenomena predictable, even in a limited way, is the existence of certain scaling properties that exhibit 'universal' behaviour. The problem of

---

[18] In earlier versions parameters like mass, charge etc. were specified at the beginning and changes in length scale simply changed the values from the bare values appearing in the basic Hamiltonian to renormalized values. The old renormalization theory was a mathematical technique used to rid quantum electrodynamics of divergences but involved no "physics".

[19] The equivalence of power laws with a particular scaling exponent can have a deeper origin in the dynamical processes that generate the power-law relation. Phase transitions in thermodynamic systems are associated with the emergence of power-law distributions of certain quantities, whose exponents are referred to as the critical exponents of the system. Diverse systems with the same critical exponents—those that display identical scaling behaviour as they approach criticality—can be shown, via RG, to share the same fundamental dynamics.

calculating the critical indices for these different systems was impossible prior to the use of renormalization group techniques which enable us to see that different kinds of transitions such as liquid-gas, magnetic, alloy etc. share the same critical exponents and can be understood in terms of the same fixed-point interaction.

As I noted above, epistemic independence—the fact that we *need not* appeal to micro phenomena to explain macro processes—is not sufficient for emergence since it is also a common feature of physical explanation across many systems and levels. Emergence is characterized by the fact that we *cannot* appeal to microstructures in explaining or predicting these phenomena despite their microphysical base. RG methods reveal the nature of this ontological independence by demonstrating the features of universality and how successive transformations give you a Hamiltonian for an ensemble that contains very different couplings from those that governed the initial ensemble.

Despite the explanatory power of fixed points, Butterfield (2011) has recently claimed that one needn't resort to RG in explaining phase transitions. Indeed there is a sense in which this is true if what we are trying to explain is the appearance of stable behaviour in finite systems; the sort of behaviour that we sometimes identify with phase transitions (e.g. the appearance of critical opalescence). Many (e.g. Callender 2001; Earman 2004) have argued that appeals to infinite systems required to explain phase transitions is, in fact, illegitimate since we know that the relevant behaviour occurs in finite systems. Issues related to the stability of finite system behaviour has also been pointed out by Menon and Callender (2013) and well as Huttemann, Kuhn and Terzidis (this volume). In each of these cases, however, the authors ignore a crucial feature of emergence, specifically the ability to properly explain universal behaviour and, in Butterfield's case, the role of RG in that context. The calculation of values for critical indices and the cooperative behaviour defined in terms of fixed points is the foundation of universality. RG is the only means possible for explaining that behaviour; what happens at finite N is, in many ways, irrelevant. Finite systems can be near the fixed point in the RG space and linearization around a fixed point will certainly tell you about finite systems, but the fixed point itself requires the limit.

What RG does is show us how to pass through the various scales to reach the point where phase transitions are not breakdowns in approximation techniques, but true physical effects. We know that if you try to approximate a sum by an integral you quickly find that exact summation can't admit a phase transition. And, although we witness stable and universal behaviour experimentally in finite N, we aren't able to understand its fundamental features without RG. The formal (mathematical) features function as indicators of the kind of phenomena we identify with phase transitions and in that sense the mathematics provides a representation and precise meaning for the relation between phase transitions and universal behaviour.

Many of the worries surrounding emergence are related to the issue of reduction and whether the former presents a telling case against the latter. Why, for example, should universality be considered more effective against reduction than multiple realizability arguments? Moreover, one could also claim that universality and symmetry breaking are part of fundamental physics and hence the emergentist story actually encorporates elements of reduction. The objection concerning symmetry breaking and fundamental physics can be answered as follows: Although what defines fundamental physics is not rigidly designated it unequivocally includes explanations that invoke microphysical entities and theories/laws that govern then. When symmetry breaking features in microphysical theories its role requires specific details of the 'breaking', i.e. an account that appeals to microstructures as in the case of the Higgs mechanism. The point of the superconductivity example was to illustrate that details of symmetry breaking were not necessary for the derivation of infinite conductivity; all that was required was an assumption that electromagnetic gauge invariance was spontaneously broken. So, while SSB bears some relation to microphysical explanation, as a general process it doesn't qualify as "fundamental" in the way the term is typically understood. The existence of universal phenomena further bears this out. Because we witness identical behaviour at critical point from phenomena that have completely different microstructures, and the explanation ignores those microstructures, the notion of fundamental physics is rendered inapplicable.

Here the reductionist might respond that surely it is possible *in principle* to derive macro phenomena from micro properties given the Schrodinger equation and the appropriate initial conditions (i.e. god could do it). But again, universality speaks against this possibility. If we suppose that micro properties could determine macro properties in cases of emergence then we have no explanation of how universal phenomena are even possible. Because the latter originate from vastly different micro properties there is no obvious ontological or explanatory link between the micro-structure and macro behaviour. More specifically, while fluids and magnets both arise out of microphysical constituents their behavioural similarity at criticality is independent of and immune from changes in those micro constituents. This is what separates emergent phenomena from resultant properties and aggregates. In the latter cases there is a direct physical link between the micro and macro that is absent in cases of emergence.

A relatively similar point can be made for cases of multiple realizability. Although macroregularities can be realized by radically heterogeneous lower level mechanisms the problem here is one of underdetermination; we simply don't know which of the micro arrangements is responsible for the macro state and hence the causal, explanatory link is unknown with respect to the competing alternatives. However, universality presents a rather different picture in that the micro-macro link is simply broken rather than being underdetermined. In other words, we know what the initial macro states are in each of the separate instances of critical behaviour, but because those are "washed out" after several iterations of RG equations they no longer play a role in the macro behaviour. Moreover, the mystery to be explained is how several different systems with different micro structures behave in exactly the same way; hence, because the micro structures are different in each case the explanation *cannot* be given in those terms. In that sense the analogy with multiple realizability breaks down.

A more direct challenge to the claim about the incompatibility of emergence and reduction comes from Huttemann et.al. (this volume). They claim that "the fact that certain features of the constituents are irrelevant in the technical RG sense does not imply that the properties and states of the constituents fail to influence macro-behaviour. Rather, it is only a small number of features of these that does the work for asymptotic critical exponents." But these features are simply the symmetry and dimensionality of the system and have nothing to do with claims about micro-reduction.

Finally one might want to claim that universality is simply another form of multiple realizability (MR) and to that extent is provides no added reason to deny reduction and embrace emergence. The possibility that macro-level regularities are heterogeneously multiply-realised is evidenced by the fact that liquids and gases exhibit the same type of behaviour at critical point while having radically different microstructures. So, the issue is whether examples of universal behaviour fall prey to some form of reduction in virtue of a supervenient relation to their microphysical base.[20]

Here again the answer is 'no'. The dependence relation required for superve-nience is clearly lacking in cases of 'universal' behaviour since fixing the subve-nient properties in no way fixes the supervenient ones and vice versa—the whole is substantially different from the sum of its parts. In cases of supervenience any change in higher level properties requires a difference in lower level properties, something that fails to occur in cases of emergence. For example, superconducting metals that constitute different "natural kinds" will have different transition tem-peratures but they exhibit the same properties as a consequence of broken elec-tromagnetic gauge invariance. The claim so often associated with supervenience—there can be no A difference without a B difference (where A properties supervene on B properties)—is irrelevant here since once the system reaches critical point and universal behaviour (A properties) is dominant, information about micro-level structure (B properties) is simply lost.

But as I have stressed many times, the issue is not simply a matter of ignoring irrelevant details as one does in the formulation of laws or levels of explanation. In those cases changes in macro structure *are* determined by changes in micro structure and vice versa. In emergence the important *physical* relationships involve long wavelengths and cooperative behaviour defined in terms of fixed points. The systematic treatment provided by RG enables us to see behind the abstract math-ematics of the thermodynamic limit and divergence of the correlation length to fully illustrate the physical processes involved in emergent 'universal' phenomena.

---

[20] Although there are arguments for the claim that supervenience needn't entail reduction my argument rests on the fact that even the requirements of supervenience fails in the case of universal phenomena.

6.6 Conclusions

One of the fundamental issues in debates about emergence involves the difference between epistemic and ontological claims about what constitutes emergent phenomena. The temptation to classify everything as epistemically emergent is overwhelming, especially due to uncertainties about what future physics will reveal. For example, it seems reasonable to suggest that our inability to explain or predict phenomena we now classify as emergent will or can be resolved once a more comprehensive theory is in place. However, once we focus on the notion of universality the appeal of epistemic emergence quickly fades. For instance, the fact that phenomena as different as liquids and magnets exhibit the same critical behaviour and share the same values for critical exponents is not going to be explained by a more comprehensive micro theory. In fact, the difference in the micro structure of phenomena that share the same universality class indicates that the explanation of their stable, emergent behaviour cannot not arise from the microphysical base. In that sense universality undermines any appeals to reduction as an explanatory strategy for understanding this behaviour.

While emergent phenomena may be novel and surprising, these are not the characteristics by which they should be defined. Instead we need to focus on the ontological aspects of these phenomena to understand not only the basis for their similarity but also the stability of their behaviour patterns. The success of renormalization group methods in calculating the values of critical indices as well as exposing the reasons behind the failure of mean field theory in explaining universality further indicates the irrelevance of micro level, reductive explanations. However, it isn't simply the irrelevance of micro structures that it important here but also the way in which fixed points account for the cooperative behaviour present in cases of emergence. Without the explanation of these physical features via RG methods, emergent phenomena would remain theoretical novelties awaiting explanation in terms of some future theory.

References

Anderson, P.W.: More is different. Science **177**, 393–396 (1972)

Bangu, S.: Understanding thermodynamic singularities: phase transitions, data, and phenomena. Philos. Sci. **76**, 488–505 (2009)

Bardeen, J., Cooper, L.N., Schrieffer, J.R.: Theory of superconductivity. Phys. Rev. **108**, 1175–1204 (1957)

Batterman, R.: The Devil in the Details: Asymptotic Reasoning in Explanation, Reduction and Emergence. Oxford University Press, Oxford (2002)

Batterman, R.: Emergence, singularities and symmetry breaking. Found. Phys. **41**, 1031–1050 (2011)

Beckermann, A.: Supervenience, emergence and reduction. In: Beckerman, A., Flohr, H., Kim, J. (eds.) Emergence or Reduction? pp. 94–118. deGruyter, Berlin (1992)

Butterfield, J.: Less is different: emergence and reduction reconciled. Found. Phys. **41**, 1065–1135 (2011)

Callender, C.: Taking thermodynamics too seriously. Stud. Hist. Philos. Mod. Phys. **32**, 539–553 (2001)

Davidson, D.: Mental events. In: Foster, L., Swanson, J.W. (eds.) Experience and Theory, pp. 79–101. University of Massachusetts Press, Amherst (1970)

Earman, J.: Curie's principle and spontaneous symmetry breaking. Int. Stud. Philos. Sci. **18**, 173–198 (2004)

Elitzur, S.: Impossibility of spontaneously breaking local symmetries. Phys. Rev. D **12**, 3978–3982 (1975)

Friedrich, S.: Gauge symmetry breaking in gauge theories—in search of clarification. Eur. J. Philos. Sci. **3**, 157–182 (2013)

Frolich, J., Morchio, G., Strocchi, F.: Nucl. Phys. B **190,** 553 (1981)

Howard, D.: Reduction and emergence in the physical sciences: some lessons from the particle physics-condensed matter physics debate. In: Nancey, M., Willian, R., Stoeger, S.J. (eds.) Evolution, Organisms and Persons. Oxford University Press, Oxford (2007)

Humphreys, P.: How properties emerge. Philos. Sci. **64**, 1–17 (1997a)

Humphreys, P.: Emergence not supervenience. Philos. Sci. **64** (Proceedings), S337–S345 (1997b)

Laughlin, R.B., Pines, D.: The theory of everything. Proc. Nat. Acad Sci. **97**, 28–31 (2000)

Menon, T., Callender, C.: Turn and face the strange... Ch-ch-changes: philosophical questions raised by phase transitions. In: Robert Batterman (ed.) The Oxford Handbook of Philosophy of Physics, pp. 189–223. Oxford University Press (2013)

Morrison, M.: Where have all the theories gone? Philos. Sci. **74**, 195–227 (2007)

Morrison, M.: Models as representational structures. In: Bovens, L., Hartmann, Stephan, Hoefer, Carl (eds.) The Philosophy of Nancy Cartwright, pp. 67–88. Routledge, London (2008)

Rueger, A.: Physical emergence, diachronic and synchronic. Synthese **124**, 297–322 (2000)

Silberstein, M., McGeever, J.: The search for ontological emergence. Philos. Q. **49**, 182–200 (1999)

Wilson, K.G.: The renormalization group and critical phenomena. Rev. Mod. Phys. 47, 583–600 (1983)

Weinberg, S.: Superconductivity for particular theorists. Prog. Theor. Phys. Suppl. **86**, 43–53 (1986)

Weinberg, S.: The Quantum Theory of Fields. vol. II Modern Applications. Cambridge University Press, Cambridge (1996)

Chapter 7
Autonomy and Scales

Robert Batterman

7.1 Introduction

One way of understanding the nature of emergence in physics is by contrasting it with the reductionist paradigm that is so prevalent in high energy or "fundamental" physics. This paradigm has been tremendously successful in explaining and describing various deep features of the universe. The goal is, ultimately, to search for the basic building blocks of the universe and then, having found them, provide an account of the nonfundamental features of the world that we see at scales much larger than those investigated by particle accelerators. On this way of thinking, emergent phenomena, if there are any, are those that apparently are not reducible to, or explainable in terms of, the properties and behaviors of these fundamental building blocks. The very talk of "building blocks" and fundamental particles carries with it a particular, and widespread view of how to understand emergence in contrast with reductionism: In particular, it strongly suggests a mereological or part/whole conception of the distinction.[1] Emergent phenomena, on this conception, are properties of systems that are novel, new, unexplainable, or unpredictable in terms of the *components or parts* out of which those systems are formed. Put crudely, but suggestively, emergent phenomena reflect the possibility that the whole can be greater (in some sense) that the sum of its parts.

While I believe that sometimes one can think of reduction in contrast to emergence in mereological terms, in many instances the part/whole conception misses

---

[1] Without doing a literature survey, as it is well-trodden territory, one can simply note that virtually every view of emergent properties canvassed in O'Connor's and Wong's *Stanford Encyclopedia* article reflects some conception of a hierarchy of levels characterized by aggregation of parts to form new wholes organized out of those parts (O'Connor and Wong 2012).

R. Batterman (✉)
1028-A Cathedral of Learning, Pittsburgh, PA 15260, USA
e-mail: rbatterm@pitt.edu

© Springer-Verlag Berlin Heidelberg 2015   115
B. Falkenburg and M. Morrison (eds.), *Why More Is Different*,
The Frontiers Collection, DOI 10.1007/978-3-662-43911-1_7

what is actually most important. Often it is very difficult to identify what are the fundamental parts. Often it is even more difficult to see how the properties of those parts, should one be able to identify them, play a role in determining the behavior of systems at scales much larger than the length and energy scales characteristic of those parts. In fact, what is most often crucial to the investigation of the models and theories that characterize systems is the fact that there is an enormous separation of scales at which one wishes to model or understand the systems' behaviors—scale often matters, parts not so much.[2]

In this paper I am not going to be too concerned with mereology. There is another feature of emergent phenomena that is, to my mind, not given sufficient attention in the literature. Emergent phenomena exhibit a particular kind of *autonomy*. It is the goal of this paper to investigate this notion and having done so to draw some conclusions about the emergence/reduction debate. Ultimately, I will be arguing that the usual characterizations of emergence in physics are misguided because they focus on the *wrong* questions.

In the next section, I will discuss the nature and evidence for the relative autonomy of the behaviors of systems at continuum scales from the details of the systems at much lower scales. I will argue that materials display a kind of universality at macroscales and that this universal behavior is governed by laws of continuum mechanics of a fairly simple nature. I will also stress the fact (as I see it) that typical philosophical responses to what accounts for the autonomy and the universality—both from purely bottom-up and a purely top-down perspectives—really miss the subtleties involved. Section 7.3 will discuss in some detail a particular strategy for bridging models of material behavior across scales. The idea is to replace a complicated problem that is heterogeneous at lower scales with an effective or homogeneous problem to which the continuum equations of mechanics can then be applied. Finally, in the conclusion I will draw some lessons from these discussions about what are fruitful and unfruitful ways of framing the debate between emergentists and reductionists.

7.2 Autonomy

References to the term "emergence" in the contemporary physics literature and in popular science sometimes speak of "protected states of matter" or "protectorates." For example, Laughlin and Pines (2007) use the term "protectorate" to describe domains of physics (states of matter) that are effectively independent of microdetails of high energy/short distance scales. A (quantum) protectorate, according to Laughlin and Pines is "a stable state of matter whose generic low-energy properties

---

[2] Some examples, particularly from theories of optics, where one can speak of relations between theories and models where no part/whole relations seem to be relevant can be found in Batterman (2002). Furthermore, it is worth mentioning that there are different kinds of models than may apply at the same scale from which one might learn very different things about the system.

are determined by a higher organizing principle and nothing else" (Lauglin and Pines 2007, p. 261). Laughlin and Pines do not say much about what a "higher organizing principle" actually is, though they do refer to spontaneous symmetry breaking in this context. For instance, they consider the existence of sound in a solid as an emergent phenomenon independent of microscopic details:

> It is rather obvious that one does not need to prove the existence of sound in a solid, for it follows from the existence of elastic moduli at long length scales, which in turn follows from the spontaneous breaking of translational and rotation symmetry characteristic of the crystalline state" (Lauglin and Pines 2007, p. 261).

It is important to note that Laughlin and Pines do refer to features that exist at "long length scales." Unfortunately, both the conception of a "higher organizing principle" and what they mean by "follows from" are completely unanalyzed. Nevertheless, these authors do seem to recognize an important feature of emergent phenomena—namely, that they display a kind of autonomy. Details from higher energy (shorter length) scales appear to be largely irrelevant for the emergent behavior at lower energies (larger lengths).

7.2.1 Empirical Evidence

What evidence is there to support the claim that certain behaviors of a system at a given scale are relatively autonomous from details of that system at some other (typically smaller or shorter) scale? In engineering contexts, materials scientists often want to determine what kind of material is appropriate for building certain kinds of structures. Table 7.1 lists various classes of engineering materials. If we want to construct a column to support some sort of structure, then we will want to know how different materials will respond to loads with only minimal shear forces. We might be concerned with trade-offs between weight (or bulkiness), price of materials, etc. We need a way of comparing possible materials to be used in the construction. There are several properties of materials to which we can appeal in making our decisions, most of which were discovered in the nineteenth century. Young's modulus, for instance, tells us how a material changes under tensile or compressive loads. Poisson's ratio would also be relevant to our column. It tells us how a material changes in directions perpendicular to the load. For instance if we weight our column, we will want to know how much it fattens under that squeezing. Another important quantity will be the density of the material. Wood, for example, is less dense than steel. Values for these, and other parameters have been determined experimentally over many years. It is possible to classify groups of materials in terms of trade-offs between these various parameters.

Michael Ashby has developed what he calls "Material Property Charts" with which it is possible to plot these different properties against one another and exhibit quite perspicuously what range of values fit the various classes of materials exhibited in Table 7.1. An illustrative example is provided in Fig. 7.1. There are a

Table 7.1 Classes of materials (after Ashby (1989))

Engineering alloys	Metals and their alloys
Engineering polymers	Thermoplastics and thermosets
Engineering ceramics	"Fine" ceramics
Engineering composites	Carbon fibre, glass fibre, kevlar fibre
Porous ceramics	Brick, cement, concrete, stone
Glasses	Silicate glasses
Woods	Common structural timbers
Elastomers	Natural and artificial rubbers
Foams	Foamed polymers

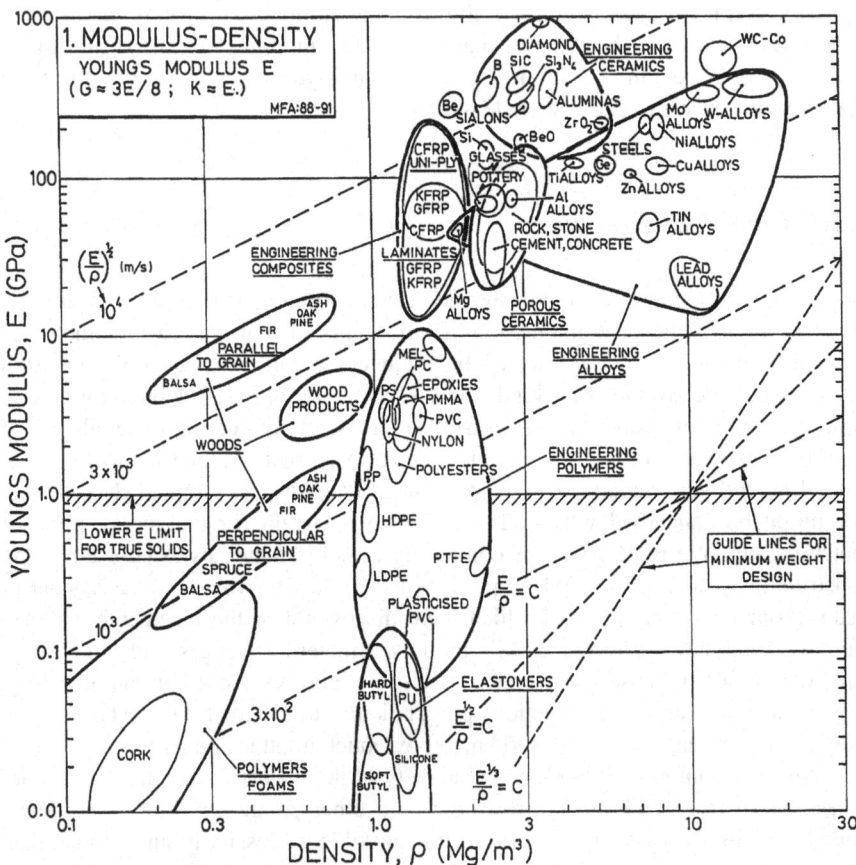

Fig. 7.1 Density versus young's modulus

number of things to note about this chart. First, it is plotted in a log scale. This is for two reasons. On the one hand it allows one to represent a very large range of values for the material properties like density (ρ) and Young's modulus (E). In a number of instances these values, for a given class of materials, can range over 5 decades. On the other hand, the log plot allows one to represent as straight lines the longitudinal elastic wave velocity—the speed of sound in the solid. These are the parallel lines in units of $\left(\frac{E}{\rho}\right)^{1/2}$ (m/s). Clearly the wave velocity is greater in steel and other engineering alloys than it is in cork or in elastomers.

One can also plot the contours E/ρ, $E^{1/2}/\rho$, and $E^{1/3}\rho$ that respectively tell us how a tie rod, column, and plate or panel will deflect under loading. For example,

[t]he lightest tie rod which will carry a given tensile load without exceeding a given deflection is that with the greatest value of E/ρ. The lightest column which will support a given compressive load without buckling is that with the greatest value of $E^{1/2}/\rho$. The lightest panel which will support a given pressure with minimum deflection is that with the greatest value of $E^{1/3}/\rho$.

On the chart these contours are marked as "Guide Lines For Minimum Weight Design." In order to use them one parallel shifts the line toward the upper left hand corner. The intersection of the line with different materials gives you the materials that will support the same load with different densities. To save weight, choose that material with the smallest density that intersects the line. Thus, perpendicular to the grain woods will support a column as well as steel and its various alloys—they will have equal values of $E^{1/2}/\rho$.

This chart, and others like it, provide a qualitative picture of both the mechanical and thermal properties of various engineering materials. For our purposes, the most interesting feature is how different materials can be grouped into classes. Within a given envelope defining an entire class of materials from Table 7.1 (bold lines in Fig. 7.1) one can also group individual instances into balloons (lighter lines) that represent particular types of materials in those general classes. These display materials that exhibit similar behaviors with respect to the parameters of interest. Ashby sums this up nicely:

The most striking feature of the charts is the way in which members of a material class cluster together. Despite the wide range of modulus and density associated with metals (as an example), they occupy a field which is distinct from that of polymers, or that of ceramics, or that of composites. The same is true of strength, toughness, thermal conductivity and the rest: the fields sometimes overlap, but they always have a characteristic place within the whole picture (Ashby 1989, p. 1292).

An immediate question concerns why the materials group into distinct envelopes in this way. The answer must have something to do with the nature of the atomic make-up of the materials. For example, a rough, but informative, answer to why materials have the densities they do, depends upon the mean atomic weight of the atoms, the mean size of the atoms and the geometry of how they are packed.

> Metals are dense because they are made of heavy atoms, packed more or less closely;
> polymers have low densities because they are made of light atoms in a linear, 2 or
> 3-dimensional network. Ceramics, for the most part, have lower densities than metals
> because they contain light O, N or C atoms....
> The *moduli* of most materials depend on two factors: bond stiffness and the density of
> bonds per unit area. A bond is like a spring.... (Ashby 1989, pp. 1278–1279).

This provides a justification for the observed fact that distinct materials can be grouped into (largely non-overlapping) envelopes in the material property charts, but it does not fully account for the shapes of the envelopes. Why, for instance, are some envelopes elongated in one direction and not another? To answer questions of this sort, we need to pay attention to structures that exist at scales in between the atomic and the macroscopic. The importance of these "mesoscale" structures will be the primary concern of this paper.

This section was meant simply to exhibit the fact that the behaviors of materials have been empirically determined to lie within certain classes that are defined in terms of material parameters whose importance was noted already in the nineteenth century. These so-called moduli are defined at the macroscale where materials are represented as a continuum. The fact that the materials group into envelopes in the property charts is an empirically determined fact that characterizes types of *universal* behavior for different engineering materials. That is to say, the atomically and molecularly different materials within the same envelope behave, at the macroscale, in very similar ways. A question of interest is how this universal behavior can be understood. Put in a slightly different way: Can we understand how different materials come to be grouped in these envelopes with their distinct shapes? There seem to be "organizing principles" at play here. Can we say anything about these macroscopic moduli beyond simply asserting that they exists?[3]

7.2.2 The Philosophical Landscape

As we have just seen, the material property charts provide plenty of evidence of universal behavior at macroscales. This is also evidence that the behaviors of different materials at those scales are relatively autonomous from the details of the materials at atomic scales. Furthermore, the equations of continuum mechanics exhibit a kind of safety: They enable us to safely build bridges, boats, and airplanes.

The question of how this safety, autonomy, and universality are possible has received virtually no attention in the philosophical literature.[4] I think the little that has been said can be seen as arising from two camps. Roughly, these correspond to the emergentist and reductionist/fundamentalist poles in the debate. Emergentists

---

[3] See (Lauglin and Pines 2007, p. 261), quoted above.

[4] Exceptions appear in some discussions of universality in terms of the renormalization group theory of critical phenomena and in quantum field theory. However, universality, etc., is ubiquitous in nature and more attention needs to be paid to our understanding of it.

might argue that the very simplicity of the continuum laws can be taken to justify the continuum principles in a sense as "special sciences" independent of "lower-level" details. Recall Fodor's, essentially all or nothing understanding of the autonomy of the special sciences:

> [T]here are special sciences not because of the nature of our epistemic relation to the world, but because of the way the world is put together: not all natural kinds (not all the classes of things and events about which there are important, counterfactual supporting generalizations to make) are, or correspond to, physical natural kinds.... Why, in short, should not the natural kind predicates of the special sciences cross-classify the physical natural kinds? (Fodor 1974, pp. 113–114)

I think this is much too crude. It is, in effect, a statement that the special sciences (in our case the continuum mechanics of materials) are *completely autonomous* from lower scale details. But we have already seen that the density of a material will depend in part on the packing of the atoms in a lattice. Furthermore, on the basis of continuum considerations alone, we cannot hope to understand which materials will actually be instantiated in nature. That is, we cannot hope to understand the groupings of materials into envelopes as in the property charts; nor can we hope to understand why there are regions within those envelopes where no materials exist. From a design and engineering point of view, these are, of course, key questions.

The shortcomings of a fully top-down approach may very well have led some fundamentalists or reductionists to orthogonal prescriptions about how to understand and explain the evidence of universality and relative autonomy at macroscales. They might argue as follows: "Note that there are roughly one hundred elements in the periodic table rather than the potential infinity of materials that are allowed by the continuum equations. So, *work at the atomic level!* Start with atomic models of the lattice structure of materials and determine, on the basis of these fundamental features the macroscale properties of the materials for use in engineering contexts."

I think this is also much too crude. The idea that one can, essentially ab initio, bridge across 10+ orders of magnitude is a reductionist pipe-dream. Consider the following passage from an NSF Blue Ribbon report on simulation based engineering science.[5]

> Virtually all simulation methods known at the beginning of the twenty-first century were valid only for limited ranges of spatial and temporal scales. Those conventional methods, however, cannot cope with physical phenomena operating across large ranges of scale—12 orders of magnitude in time scales, such as in the modeling of protein folding... or 10 orders of magnitude in spatial scales, such as in the design of advanced materials. At those ranges, the power of the tyranny of scales renders useless virtually all conventional methods.... Confounding matters further, the principal physics governing events often changes with scale, so that the models themselves must change in structure as the ramifications of events pass from one scale to another.
>
> The tyranny of scales dominates simulation efforts not just at the atomistic or molecular levels, but wherever large disparities in spatial and temporal scales are encountered (Oden 2006, pp. 29–30).

---

[5] See Batterman (2013) for further discussion.

While this passage refers to simulations, the problem of justifying simulation techniques across wide scale separations, is intimately connected with the problem of interest here—justifying the validity of modeling equations at wide scale separations. The point is that more subtle methods are required to explain and understand the relative autonomy and universality observed in the behavior of materials at continuum scales. The next section discusses one kind of approach to bridging scales that involves both top-down and bottom-up modeling strategies.

7.3 Homogenization: A Means for Upscaling

Materials of various types appear to be homogeneous at large scales. For example if we look at a steel beam with our eyes it looks relatively uniform in structure. If we zoom in with low powered microscopes we see essentially the same uniform structure. (See Fig. 7.2.) The material parameters or moduli characterize the behaviour of these homogeneous structures. Continuum mechanics describes the behaviors of materials at these large scales assuming that the material distribution, the strains, and stresses can be treated as homogeneous or uniform. Material points and their neighborhoods are taken to be uniform with respect to these features. However, if we zoom in further, using x-ray diffraction techniques for example, we begin to see that the steel beam is actually quite heterogeneous in its makeup.

We want to be able to determine what are the values for the empirically discovered *effective* parameters that characterize the continuum behaviors of different materials. This is the domain of what is sometimes called "micromechanics." A major goal of micromechanics is "to express in a systematic and rigorous manner the *continuum quantities* associated with an infinitesimal material neighborhood in terms of the parameters that characterize the *microstructure and properties of the microconstituents* of the material neighborhood" (Nemat-Nasser and Hori 1999, p. 11). One of the most important concepts employed in micromechanics is that of a *representative volume element* (RVE).

7.3.1 RVEs

The concept of an RVE has been around explicitly since the 1960s. It involves explicit reference to features or structures that are absent at continuum scales. An RVE for a continuum material point is defined to be

> a material volume which is statistically representative of the infinitesimal material neighborhood of that material point. The continuum material point is called a *macro-element*.

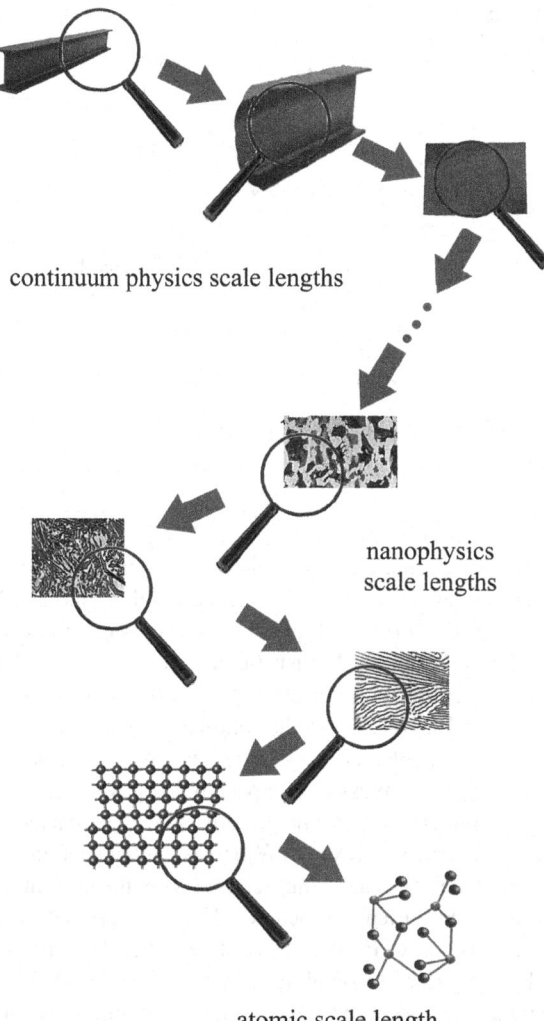

continuum physics scale lengths

nanophysics
scale lengths

atomic scale length

Fig. 7.2 Steel—widely separated scales. The corresponding microconstitutents of the RVE are called the *micro-elements*. An RVE must include a very large number of micro-elements, and be statistically representative of the local continuum properties (Nemat-Nasser and Hori 1999, p.11)

In Fig. 7.3 the point P is a material point surrounded by an infinitesimal material element. As noted, this is a *macro-element*. The inclusions, voids, cracks, and grain-boundaries are to be thought of as the *microstructure* of macro-element.

The characterization of an RVE requires (as is evident from the figure) the introduction of two length scales. There is the continuum or macro-scale (D) by which one determines the infinitesimal neighborhood of the material point, and there is a microscale (d) representing the smallest microconstituents whose properties (normally shapes) are believed to directly effect the overall response and

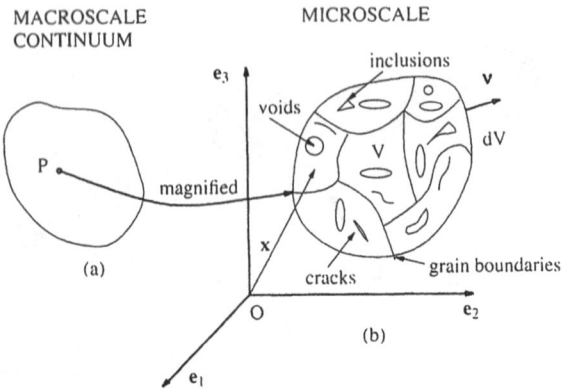

Fig. 7.3 RVE (Nemat-Nasser and Hori (1999), p. 12)

properties of the continuum infinitesimal material neighborhood. We can call these the "essential" microconstituents. These length scales must typically differ by orders of magnitude so that $d/D \ll 1$. The requirement that this ratio be very small is independent of the nature of the distribution of the microconstituents. They may, for instance, be periodically or randomly distributed throughout the RVE. Surely, what this distribution is will effect the overall properties of the RVE.

The concept of an RVE is *relative* in the following sense. The actual characteristic lengths of the microconstituents can vary tremendously. As Nemat-Nasser and Hori note, the overall properties of a mass of compacted fine powder in powder-metallurgy can have grains of micron size, so that a neighborhood with characteristic dimension of 100 microns could well serve as a RVE. "[W]hereas in characterizing a earth dam as a continuum, with aggregates of many centimetres in size, the absolute dimension of an RVE would be of the order of tens of meters" (Nemat-Nasser and Hori 1999, p. 15). Therefore, the concept of an RVE involves relative rather than absolute dimensions.

Clearly it is also important to be able to identify the "essential" microconstituents. This is largely an art informed by the results of experiments. The identification is a key component in determining what an appropriate RVE is for any particular modeling problem. Again, here is Nemat-Nasser and Hori: "An optimum choice would be one that includes the most dominant features that have first-order influence on the overall properties of interest and, at the same time, yields the simplest model. This can only be done through *a coordinated sequence* of microscopic (small-scale) and macroscopic (continuum scale) observations, experimentation, and analysis" (Nemat-Nasser and Hori 1999, p. 15, My emphasis).

The goal is to employ the RVE concept to extract the macroscopic properties of such micro-heterogeneous materials. There are several possible approaches to this in the literature. One approach is explicitly statistical. It employs n-point correlation functions that characterize the probability, say, of finding heterogeneities of the same type at n different points in the RVE. The RVE is considered as a member of an ensemble of RVEs from which macroscopic properties such as moduli and material responses to deformations are determined by various kinds of averaging techniques. For this approach see (Torquato 2002). In this paper, however, I will discuss a nonstatistical, classical approach that is described as follows:

> ... the approach begins with a simple *model*, exploits fundamental principles of continuum mechanics, especially linear elasticity and the associated extremum principles, and, estimating local quantities in an RVE in terms of global boundary data, seeks to compute the overall properties and associated *bounds* (Nemat-Nasser and Hori 1999, p. 16).

This approach is, as the passage notes, fully grounded in continuum methods. Thus, it employs *continuum* mechanics to consider the heterogeneous grainy structures and their effects on macroscale properties of materials. As noted, one goal is to account for the shapes of the envelopes (the range of values of the moduli and density) appearing in Ashby's material property charts. Thus, this approach is one in which top-down modeling is employed to set the stage for upscaling—for making (bottom-up) inferences.

7.3.2 Determining Effective Moduli

Two of the most important properties of materials at continuum scales are strength and stiffness. We will examine these properties in the regime in which we take the materials to exhibit linear elastic behavior. All materials are heterogeneous at scales below the continuum scale so we need to develop the notion of an effective or "equivalent" homogeneity. That is, we want to determine the effective material properties of an idealized homogeneous medium by taking into consideration material properties of individual phases (the various inclusions that are displayed in the RVE) and the geometries of those individual phases (Christensen 2005, pp. 33–34). (Recall that steel looks completely homogeneous when examined with our eyes and with small powered magnifiers.)

We are primarily interested in determining the behavior of a material as it is stressed or undergoes deformation. In continuum mechanics this behavior is understood in terms of stress and strain tensors which, in linear elastic theory are related to one another by a generalization of Hooke's law:

$$\sigma_{ij} = C_{ijkl}\varepsilon_{kl} \tag{7.1}$$

σ_{ij} and ε_{kl} are respectively the linear stress and strain tensors and C_{ijkl} is a fourth order tensor—the stiffness tensor—that represents Young's modulus. The latter is constitutive of the particular material under investigation. The infinitesimal strain tensor ε_{kl} is defined in terms of deformations of the material under small loads.

Assume that our material undergoes some deformation (a load is placed on a steel beam). Consider an RVE of volume V. We can define the average strain and stress, respectively as follows:

$$\langle \varepsilon_{ij} \rangle = \int\limits_V \varepsilon_{ij}(x_i)dv \qquad (7.2)$$

$$\langle \sigma_{ij} \rangle = \int\limits_V \sigma_{ij}(x_i)dv \qquad (7.3)$$

These averages are defined most generally—there are no restrictions on the boundaries between the different phases or inclusions in the RVE. Given these averages, the effective linear stiffness tensor is defined by the following relation:

$$\langle \sigma_{ij} \rangle = C_{ijkl}^{\text{eff}} \langle \varepsilon_{kl} \rangle. \qquad (7.4)$$

So the problem is now "simple." Determine the tensor, C_{ijkl}^{eff} having determined the averages (7.2) and (7.3). "Although this process sounds simple in outline, it is complicated in detail, and great care must be exercised in performing the operation indicated. To perform this operation rigorously, we need exact solutions for the stress and strain fields σ_{ij} and ε_{ij} in the heterogeneous media" (Christensen 2005, p. 35). Fortunately, it is possible to solve this problem theoretically (not empirically) with a minimum of approximations by attending to some idealized geometric models of the lower scale heterogeneities. It turns out that a few geometric configurations actually cover a wide range of types of materials: Many materials can be treated as composed of a continuous phase, called the "matrix," and a set of inclusions of a second phase that are *ellipsoidal* in nature. Limiting cases of the ellipsoids are spherical, cylindrical, and lamellar (thin disc) shapes. These limiting shapes can be used to approximate the shapes of cracks, voids, and grain boundaries.

Let us consider a material composed of two materials or phases.[6] The first is the continuous matrix phase (the white region in Fig. 7.4) and the second is the discontinuous set of discrete inclusions. We will assume that both materials are themselves isotropic. The behavior of the matrix material is specified by the Navier-Cauchy stress-strain relations

$$\sigma_{ij} = \lambda_M \delta_{ij}\varepsilon_{kk} + 2\mu_M \varepsilon_{ij}. \qquad (7.5)$$

Simlarly, the behavior of the inclusion phase is given by the following:

$$\sigma_{ij} = \lambda_I \delta_{ij}\varepsilon_{kk} + 2\mu_I \varepsilon_{ij}. \qquad (7.6)$$

The parameters λ and μ are the Lamé parameters that characterize the elastic properties of the different materials (subscripts indicate for which phase). They are related to Young's modulus and the shear modulus.

---

[6] This discussion follows (Christensen 2005, pp. 36–37).

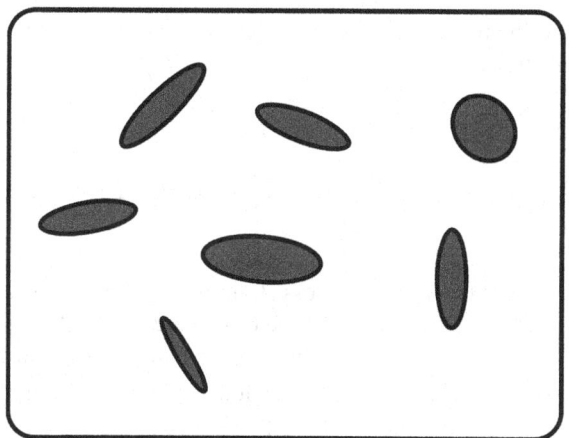

Fig. 7.4 RVE with ellipsoidal inclusions

We can now rewrite the average stress formula (7.3) as follows:

$$\langle \sigma_{ij} \rangle = \frac{1}{V} \int\limits_{V - \sum\limits_{n=1}^{N} V_n} \sigma_{ij} dv + \frac{1}{V} \sum_{n=1}^{N} \int\limits_{V_n} \sigma_{ij} dv. \tag{7.7}$$

Thus, there are N inclusions with volumes V_n and $V - \sum_{n=1}^{N} V_n$ represents the total volume of the matrix region. Now we use the Navier-Cauchy Eq. (7.5) to get the following expression for the average stress:

$$\langle \sigma_{ij} \rangle = \frac{1}{V} \int\limits_{V - \sum\limits_{n=1}^{N} V_n} \left(\lambda_M \delta_{ij} \varepsilon_{kk} + 2\mu_M \varepsilon_{ij} \right) dv + \frac{1}{V} \sum_{n=1}^{N} \int\limits_{V_n} \sigma_{ij} dv. \tag{7.8}$$

The first integral here can be rewritten as two integrals (changing the domain of integration) to get:

$$\langle \sigma_{ij} \rangle = \frac{1}{V} \int\limits_{V} \left(\lambda_M \delta_{ij} \varepsilon_{kk} + 2\mu_M \varepsilon_{ij} \right) - \frac{1}{V} \sum_{n=1}^{N} \int\limits_{V_n} \left(\lambda_M \delta_{ij} \varepsilon_{kk} + 2\mu_M \varepsilon_{ij} \right) dv$$

$$+ \frac{1}{V} \sum_{n=1}^{N} \int\limits_{V_n} \sigma_{ij} dv. \tag{7.9}$$

Using Eq. (7.4) the first integral in (7.9)—the average stress—can be rewritten in terms of average strains:

$$C_{ijkl}^{\text{eff}} \langle \varepsilon_{kl} \rangle = \lambda_M \delta_{ij} \langle \varepsilon_{kk} \rangle + 2\mu_M \langle \varepsilon_{ij} \rangle$$
$$+ \frac{1}{V} \sum_{n=1}^{N} \int_{V_n} (\sigma_{ij} - \lambda_M \delta_{ij} \varepsilon_{kk} - 2\mu_M \varepsilon_{ij}) dv. \tag{7.10}$$

So we have arrived at an *effective* stress/strain relation (Navier-Cauchy equation) in which "only the conditions within the inclusions are needed for the evaluation of the effective properties tensor $[C_{ijkl}^{\text{eff}}]$" (Christensen 2005, p. 37). This, together with the method to be discussed in the next section will give us a means for finding homogeneous effective replacements for heterogenous materials.

7.3.3 Eshelby's Method

So our problem is one of trying to determine the stress, strain, and displacement fields for a material consisting of a matrix and a set of inclusions with possibly different elastic properties than those found in the matrix. The idea is that we can then model the large scale behavior *as if* it were homogeneous, having taken into consideration the microstructures in the RVE.

We first consider a linear elastic solid (the matrix) containing an inclusion of volume V_0 with boundary S_0. See Fig. 7.5. The inclusion is of the same material and is to have undergone a permanent deformation such as a (martensitic) phase transition. In general the inclusion "undergoes a change of shape and size which, but for the constraint imposed by its surroundings (the 'matrix') would be an arbitrary homogeneous strain. What is the elastic state of the inclusion and matrix?" (Eshelby 1957, p. 376) Eshelby provided an ingenious answer to this question.

Begin by cutting out the inclusion V_0 and removing it from the matrix. Upon removal, it experiences a uniform strain ε_{ij}^T with zero stress. Such a stress-free transformation strain is called an "eigenstrain" and the material undergoes an unconstrained deformation. See Fig. 7.6. Next, apply surface tractions to S_0 and restore the cut-out portion to its original shape. This yields a stress in the region V_0 as exhibited by the smaller hatch lines; but there is now *zero stress in the matrix*. Return the volume to the matrix and remove the traction forces on the boundary by apply equal and opposite body forces. The system (matrix plus inclusion) then comes to equilibrium exhibiting at a constrained strain ε_{ij}^C relative to its initial shape prior to having been removed. *Note: elastic strains in these figures are represented by changes in the grid shapes. Within the inclusions, the stresses and strains are uniform.*

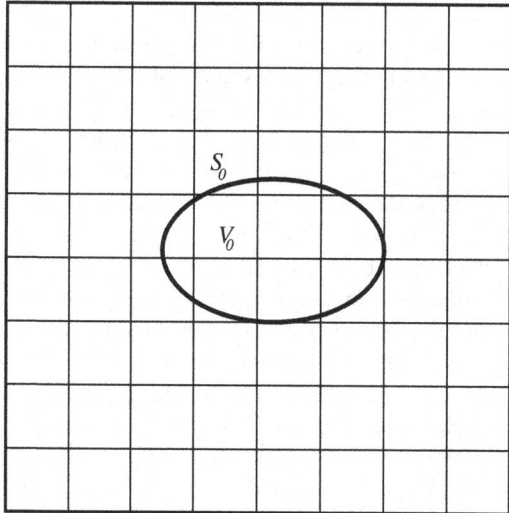

Fig. 7.5 Matrix and inclusion

By Hooke's law (Eq. (7.1)) the stress in the inclusion σ_I can be written as follows[7]:

$$\sigma_I = C_M(\varepsilon^C - \varepsilon^T).$$

Eshelby's main discovery is that if the inclusion is an ellipsoid, the stress and strain fields within the inclusion after these manipulations is *uniform* (Eshelby 1957, p. 377). Given this, the problem of determining the elastic strain everywhere is reasonably tractable. One can write the constrained strain, ε^C, in terms of the stress free eigenstrain ε^T by introducing the so-called "Eshelby tensor", S_{ijkl}:

$$\varepsilon^C = S\varepsilon^T.$$

The Eshelby tensor "relates the final constrained inclusion shape [and inclusion stress] to the original shape mismatch between the matrix and the inclusion" (Withers et al. 1989, p. 3062). The derivation of the Eshelby tensor is complicated, but it is tractable when the inclusion is an ellipsoid.

The discussion has examined an inclusion within a much larger matrix where *the inclusion is of the same material* as the matrix. A further amazing feature of Eshelby's work is that the solution to this problem actually solves much more complicated problems where, for example, the inclusion is of a different material

---

[7] This is because the inclusion is of the same material as the matrix.

and thereby *possesses different material properties*—different elastic constants. These are the standard cases of concern. We would like to be able to determine the large scale behavior of such heterogeneous materials. The Eshelby method directly extends to these cases, allowing us to find an effective homogeneous (fictitious) material with the same elastic properties as the actual heterogeneous system. (In other words, we can employ this homogenized solution as a stand in for the actual material in our design projects—in the use of the Navier-Cauchy equations.) In this context the internal region with different elastic constants will be called the "inhomogeneity" rather than the "inclusion."[8]

I will briefly describe how this works. Imagine an inhomogeneous composite with an inhomogeneity that is stiffer than the surrounding matrix. Imagine that the inhomogeneity also has a different coefficient of thermal expansion than does the matrix. Upon heating the composite, we can then expect the inhomogeneity or inclusion to be in a state of internal stress. The stiffness tensors can be written, as above, by C_M and C_I. We now perform the same cutting and welding operations we did for the last problem. The end result will appear as in the bottom half of Fig. 7.7. The important thing to note is that the matrix strains in Figs. 7.6 and 7.7 are the same.

The stress in the inhomogeneity is

$$\sigma_I = \mathbf{C}_I(\varepsilon^C - \varepsilon^{T*}). \tag{7.11}$$

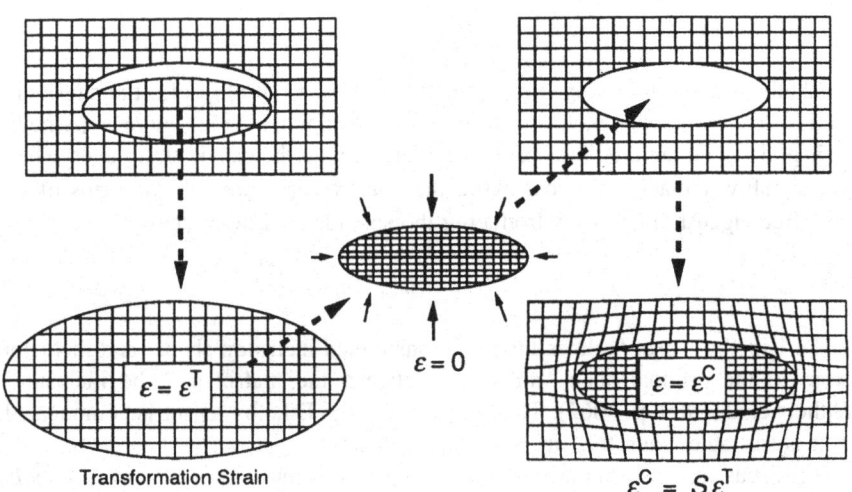

Fig. 7.6 Eshelby's method—inclusion

---

[8] Note that if the inclusion is empty, we have an instance of a porous material. Thus, these methods can be used to understand large scale behaviors of fluids in the ground. (Think of hydraulic fracturing/fracking.).

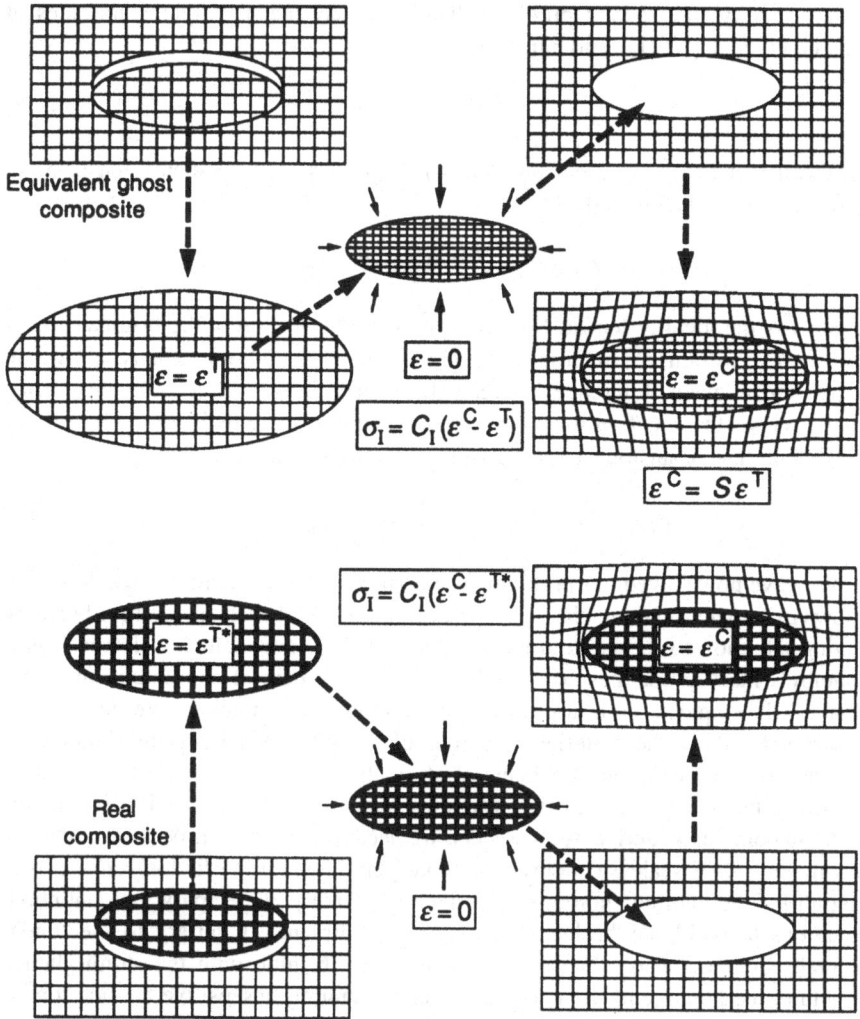

Fig. 7.7 Eshelby's method—inhomogeneity (Clyne and Withers 1993, p. 50)

The expansion of the inhomogeneity upon being cut out will in general be different than the inclusion in the earlier problem (their resulting expanded shapes will be different). This is because the eigenstrains are different $\varepsilon^{T*} \neq \varepsilon^{T}$ where ε^{T*} is the eigenstrain experienced by the inhomogeneity. Nevertheless, after the appropriate traction forces are superimposed, the shapes can be made to return to the original size of the inclusions with their internal strains equal to zero. This means that the stress in the inclusion (the first problem) must be identical to that in the

inhomogeneity (the second problem). This is to say that the right hand sides of
Eqs. (7.11) and (7.12) are identical, i.e.,

$$\sigma_I = C_M(\varepsilon^C - \varepsilon^T), \tag{7.12}$$

Recall that for the original problem $C_I = C_M$. Using the Eshelby tensor S we
now have (from (7.11) and (7.12))

$$C_I(S\varepsilon^T - \varepsilon^{T*}) = C_M(S - I)\varepsilon^T, \tag{7.13}$$

where **I** is the identity tensor. So now it is possible to express the original homogeneous eigenstrain, ε^T, in terms of the eigenstrain in the inhomogeneity, ε^{T*} which
is just the relevant difference between the matrix and the inhomogeneity. This,
finally, yields the equation for the stresses and strains in the inclusion as a result of
the differential expansion due to heating (a function of ε^{T*} alone):

$$\sigma_I = C_M(\mathbf{S} - \mathbf{I})[(C_I - C_M)S + C_M]^{-1}C_I\varepsilon^{T*}. \tag{7.14}$$

We have now solved our problem. The homogenization technique just described
allows us to determine the elastic properties of a material such as our steel beam by
paying attention to the continuum properties of inhomogeneities at lower scales.
Eshelby's method operates completely within the domain of continuum micromechanics. It employs both top-down and bottom-up strategies: We learn from
experiments about the material properties of both the matrix and the inhomogeneities. We then examine the material at smaller scales to determine the nature
(typically the geometries) of the structures inside the appropriate RVE. Furthermore, in complex materials such as steel, we need to repeat this process at a number
of different RVE scales. Eshelby's method then allows us to find *effective* elastic
moduli for the fictitious homogenized material. Finally, we employ the continuum
equations to build our bridges, buildings, and airplanes. Contrary to the funda-
mentalists/reductionists who say we should start ab initio with the atomic lattice
structure, the homogenization scheme is almost entirely insensitive to such lowest
scale structure. Instead, Eshelby's method focus largely on geometric shapes and
virtually ignores all other aspects of the heterogeneous (meso)scale structures
except for their elastic moduli. But the latter, are determined by continuum scale
measurements of the sort that figure in determining the material property charts.

7.4 Philosophical Implications

The debate between emergentists and reductionists continues in the philosophical
literature. Largely, it is framed as a contrast between extremes. Very few partici-
pants in the discussion focus on the nature or the degree of autonomy supposedly
displayed by emergent phenomena. A few, as we have seen, talk of "protected"

regions or states of matter, where this is understood in terms of an insensitivity to microscopic details. However, we have seen that to characterize the autonomy of material behaviors at continuum scales in terms of a *complete* insensitivity to microscopic details is grossly over simplified. Too much attention, I believe, has been paid to a set of ontological questions: "What is the correct ontology of a steel beam? Isn't it just completely false to speak of it as a continuum?" Of course, the steel beam consists of atoms arranged on a lattice in a very specific way. But most of the details of that arrangement are irrelevant. Instead of focusing on questions of correct ontology, I suggest a more fruitful approach is to focus on questions of proper modeling technique. Clearly models at microscales are important, but so are models at continuum scales. And, as I have been trying to argue, models at intermediate scales are probably most important of all. Importantly, none of these models are completely independent. One needs to understand the way models at different scales can be bridged. We need to understand how to pass information both upward from small scales to inform effective models at higher scales and downward to better model small scale behaviors. Despite this dependence, and also because of it, one can begin to understand the *relative autonomy* of large scale modeling from lower scale modeling. One can understand why the continuum equations work and are safe while ignoring most details at lower scales. In the context of the discussion of Eshelby's method, this is evident because we have justified the use of an effective homogenized continuum description of our (actually heterogeneous at lower scales) material.

As an aside, there has been a lot of talk about idealization and its role in the emergence/reduction debate most of which, I believe has muddied the waters.[9] Instead of invoking idealizations, we should think, for example, of continuum mechanics as it was originally developed by Navier, Green, Cauchy, Stokes, and others. Nowhere did any of these authors speak of the equations as being idealized and, hence, false for that reason. The notion of idealization is, in this context, a relative one. It is only after we had come to realize that atoms and small scale molecular structures exist that we begin to talk as though the continuum equations must be *false*. It is only then that feel we ought to be able to do everything with the "theory" that gets the ontology right. But this is simply a reductionist bias. *In fact, the continuum equations do get the ontology right at the scale at which they are designed to operate.* Dismissing such models as "mere idealization" represents a serious philosophical mistake. In the reduction/emergence debate, there has been too much focus on what is the actual fundamental level and whether, if there is a fundamental level, non-fundamental (idealized) models are dispensable. I am arguing here that the focus on the "fundamental" is just misguided.

So, the reduction/emergence debate has become mired in a pursuit of questions that, to my mind, are most unhelpful. Instead of mereology and idealization we

---

[9] I too have been somewhat guilty of contributing to what I now believe is basically a host of confusions.

should be focusing proper modeling that bridges across scales. The following quote from a primer on continuum micromechanics supports this different point of view.

> The "bridging of length scales", which constitutes the central issue of continuum micromechanics, involves two main tasks. On the one hand, the behavior at some larger length scale (the macroscale) must be estimated or bounded by using information from a smaller length scale (the microscale), i.e., homogenization or upscaling problems must be solved. The most important applications of homogenization are materials characterization, i.e., simulating the overall material response under simple loading conditions such as uniaxial tensile tests, and constitutive modeling, where the responses to general loads, load paths and loading sequences must be described. Homogenization (or coarse graining) may be interpreted as describing the behavior of a material that is inhomogeneous at some lower length scale in terms of a (fictitious) energetically equivalent, homogeneous reference material at some higher length scale. On the other hand, the local responses at the smaller length scale may be deduced from the loading conditions (and, where appropriate, from the load histories) on the larger length scale. This task, which corresponds to zooming in on the local fields in an inhomogeneous material, is referred to as localization, downscaling or fine graining. In either case the main inputs are the geometrical arrangement and the material behaviors of the constituents at the microscale (Böhm 2013, pp. 3–4).

It is clear from this passage and from the discussion of Eshelby's method of homogenization that the standard dialectic should be jettisoned in favor of a much more subtle discussion of actual methods for understanding the relative autonomy characteristic of the behavior of systems at large scales. This paper can be seen as a plea for more philosophical attention to be paid to these truly interesting methods.

References

Ashby, M.F.: On the engineering properties of materials. Acta Metall. **370**(5), 1273–1293 (1989)

Batterman, R.W.: The Devil in the Details: Asymptotic Reasoning in Explanation, Reduction, and Emergence. Oxford Studies in Philosophy of Science. Oxford University Press, Oxford (2002)

Batterman, R.W.: The tyranny of scales. In: The Oxford Handbook of Philosophy of Physics. Oxford University Press, Oxford (2013)

Böhm, H.J.: A short introduction to basic aspects of continuum micromechanics. URL http://www.ilsb.tuwien.ac.at/links/downloads/ilsbrep206.pdf (2013). May 2013

Oden, J.T.: (Chair). Simulation based engineering science—an NSF Blue Ribbon Report. URL www.nsf.gov/pubs/reports/sbes_final_report.pdf (2006)

Christensen, R.M.: Mechanics of Composite Materials. Dover, Mineola, New York (2005)

Clyne, T.W̃., Withers, P.J̃.: An Introduction to Metal Matrix Composites. Cambridge Solid State Series. Cambridge University Press, Cambridge (1993)

Eshelby, J.D.: The determination of the elastic field of an ellipsoidal inclusion, and related problems. Proc. Roy. Soc. Lond. A **2410**(1226), 376–396 (1957)

Fodor, J.: Special sciences, or the disunity of sciences as a working hypothesis. Synthese **28**, 97–115 (1974)

Lauglin, R.B., Pines, D.: The theory of everything. In: Bedau, M.A., Humphreys, P. (eds.) Emergence: Contemporary Readings in Philosophy and Science, pp. 259–268. The MIT Press, Cambridge (2007)

Nemat-Nasser, S., Hori, M.: Micromechanics: Overall Properties of Heterogeneous Materials, 2nd edn. Elsevier, North-Holland, Amsterdam (1999)

O'Connor, T., Wong, H.Y.: Emergent properties. In: Zalta E.N. (ed.) The Stanford Encyclopedia of Philosophy. Spring 2012 edn. (2012). URL http://plato.stanford.edu/archives/spr2012/entries/properties-emergent/

Torquato, S.: Random Heterogeneous Materials: Microstructure and Macroscopic Properties. Springer, New York (2002)

Withers, P.J., Stobbs, W.M., Pedersen O.B.: The application of the eshelby method of internal stress determination to short fibre metal matrix composites. Actra Metall **370**(11), 3061–3084 (1989)

Chapter 8
More is Different…Sometimes: Ising Models, Emergence, and Undecidability

Paul W. Humphreys

P. W. Anderson's paper 'More is Different' (Anderson 1972) is an iconic reference point in recent discussions of irreducibility and emergence in physics. Despite its influence, there remains considerable unclarity about what Anderson was claiming. I shall argue here for a particular interpretation of Anderson's paper, while allowing that other interpretations are possible. The principal moral of Anderson's arguments, so I shall claim, is that we must adjust the relative roles of theory and measurement within certain areas of physics and similar sciences in light of deductive limitations that are insurmountable either for theoretical or practical reasons. I shall also argue that it is premature to draw conclusions about the existence of emergent phenomena within condensed matter physics on the basis of Anderson's arguments, even though such conclusions are often made. In particular, I shall argue that formal results which sharpen some of Anderson's claims do not demonstrate the existence of emergent phenomena either within the domain of condensed matter physics itself or within the computational models used in this area.

One reason why these results are inconclusive is a product of a core problem in scientific discussions of philosophical topics. In sharp contrast to the care with which technical concepts are defined, concepts such as 'emergent', 'reducible', 'understandable', and 'global' are often used in a completely informal manner. It is therefore difficult to know exactly what it is that authors are claiming about the relation between those informal concepts and their formal results and whether authors are disagreeing with one another or are simply discussing different concepts. In the hope of improving the situation, I shall provide a number of definitions of central concepts that have appeared in this literature. Some of these definitions are uncontroversial whereas others are the subject of debate and in the foreseeable future will probably remain so.

P.W. Humphreys (✉)
Department of Philosophy, University of Virginia, 120 Cocke Hall,
Charlottesville, VA 22904, USA
e-mail: pwh2a@cms.mail.virginia.edu

© Springer-Verlag Berlin Heidelberg 2015
B. Falkenburg and M. Morrison (eds.), *Why More Is Different*,
The Frontiers Collection, DOI 10.1007/978-3-662-43911-1_8

A second reason for the inconclusiveness of these results is that, despite the common view that reducibility and emergence partition the space of possibilities, irreducibility or, in Anderson's terms, lack of constructibility, is insufficient for emergence. Additional conditions must be satisfied for emergence to obtain, many of which are either controversial or require the solution of currently open problems to assess.

8.1 Anderson's Claims

In what follows I shall indicate widely accepted technical definitions from logic and mathematics by (M), widely accepted definitions from physics by (P), widely accepted definitions from philosophy by (Phil) and definitions about which there is significant disagreement by (D), usually with an advocate's name attached. Unless enclosed in quotes, the wording of the definitions is my own.

One of Anderson's most important points was to note that the truth of a reductionist position does not entail that the converse constructionist project will be successful.

Definition (Anderson 1972, p. 393, D): A reductionist position is the view that all natural phenomena are the result of a small set of fundamental laws.

Definition (Anderson 1972, p. 393, D): A constructionist position is one asserting that it is possible in practice to construct true descriptions of all non-fundamental phenomena by starting with descriptions of fundamental laws and whatever descriptions of specific fundamental facts are needed.[1]

Anderson himself was fully committed to reduction in his sense: 'All ordinary matter obeys simple electrodynamics and quantum theory and that really covers most of what I shall discuss' (Anderson op. cit. p. 394), but, as a condensed matter physicist, he recognized that he was faced with descriptive and predictive tasks that cannot be carried out by ab initio derivational procedures starting from fundamental physics and that a constructionist position was, as a result, untenable.

It is very easy when reading Anderson's article to slide back and forth between interpreting his claims in certain passages as being about the physical phenomena themselves and in other passages as being about our theoretical descriptions of those phenomena. One reason is that whereas philosophers generally try to not conflate laws with law statements, it is common in the scientific literature to use 'laws' as a generic term that can refer to either category. Although there are places where Anderson's use of 'law' seems to favor one rather than the other of these meanings, the most plausible reading, particularly of the constructionist position, and one that maintains consistency, is to take Anderson as uniformly discussing our

---

[1] Note that this definition allows that not all fundamental laws and fundamental facts can be captured in whatever representational apparatus is used for fundamental physics, be it mathematical, computational, diagrammatic, or some other type.

theoretical representations of the physical world. I shall therefore take the reductionist position as concerning representational entities such as theories and models rather than as being directly about the systems that they represent. There is a similar, but often more subtle, slide between the ontological and the representational modes of discussion in some of the results by other authors presented below and I shall note this as appropriate.

It is common for scientists to consider the distinction between laws and law statements as a matter of 'mere semantics' but without this distinction we erase important differences between emergence as a feature of the world and emergence as a limitation on what we can predict. I briefly note here that limitations on what we can predict are different from limitations on what we can know. We can often empirically measure unpredictable values of a physical quantity. That is why I use the term 'inferential emergence' below rather than 'epistemological emergence'. Here are two definitions that separate two different types of emergence:

Definition (Phil): Ontological emergence is the position that emergent features exist as part of the natural world.

Definition (D) Inferential emergence with respect to science S is the position that it is either not possible in practice or it is impossible in principle to derive some sentences about the subject matter of S from sentences about the subject matter of sciences that are more fundamental then S and that such underivable sentences describe features of the world that count as emergent.

Here a science T is more fundamental than a science S just in case S occurs higher than T in the standard hierarchy of sciences. (See Anderson op. cit. p. 393)

I note here a primary reason that the definition of inferential emergence is flagged with a 'D': it fails to include features that are commonly required of emergence, such as that emergent features be novel and have some holistic aspect. This is one reason why, once Anderson's claims are restricted to the representational realm, the failure of constructionism is by itself not obviously sufficient for emergence. This issue will be discussed in greater detail below

Ontological emergence should be separated from what is often called 'strong emergence':

Definition (Phil): Strong emergence is a type of ontological emergence in which downwards causation from higher level property instances to lower level property instances occurs.

Strong emergence and ontological emergence break apart in areas such as fusion emergence (Humphreys 1997), an approach to ontological emergence that is not committed to downward causal relations. For this reason I shall not use the concept of strong emergence here.

Although he was discussing examples that are often cited as cases of emergence, such as broken symmetry, phase transitions, and the like, the word 'emergent' and related terms do not occur in Anderson's (1972) article.[2] Indeed, the sub-title of Anderson's article is 'Broken symmetry and the nature of the hierarchical structure

---

[2] They are used in some of Anderson's later writings, such as Anderson and Stein (1984).

of science' and one can make a case that broken symmetry and the $N \to \infty$ limit rather than emergence are Anderson's real concerns. However, Anderson's article is usually taken to be explicitly making a case for the non-constructibility of parts of condensed matter physics from high energy physics, and implicitly for the existence of emergent phenomena in the former, and so we shall have to assess the evidence for the truth of this claim, primarily by examining the work of other authors who have taken Anderson's anti-constructionist position as their starting point.

8.2 Undecidability Results

In an article directly addressing Anderson's claims (Gu et al. 2009) prove a result that they take to establish the existence of emergent features in Ising models. Although they do not give a definition for the sense of 'emergent' that they use, they appear to be using it in a way that is common in the computational condensed matter community:

Definition (Gu et al. 2009 *passim*, D): A property P of a system S is emergent just in case P is a macroscopic property and there is at least one sentence describing a value of P that is undecidable with respect to a theory of the properties of, and interactions between, the fundamental micro-constituents of S together with a specification of S's initial conditions.

This definition is consistent with our definition of inferential emergence when the characteristically emergent feature is that of a macroscopic property:

Definition (Gu et al. 2009, Sect. V, D): A macroscopic property of a physical system S that has $e_1,\ldots,e_n$ as basic components is a property of S taken as a whole and not of any of the e_i.[3]

Microscopic properties in contrast are properties of the components and not of the whole.

This requirement of macroscopic properties reflects the traditional emphasis on holistic features in emergentism.

In the definition of an emergent property, the reference to undecidable values of observables is based on the following standard definition:

Definition (M): A sentence S of a theory T is undecidable (with respect to T) just in case neither S nor not-S is a theorem of T.

An appeal to undecidability in the characterization of emergence is a way of making precise what 'in principle unpredictable' means within the emergentist tradition that identifies emergence with a certain type of unavoidable unpredictability. Put bluntly, if S is a sentence describing some state of a system **S**, on the

---

[3] A different and perhaps more precise definition of 'macroscopic' is: A macroscopic system is one whose equation of state is independent of its size and an ideal macroscopic system is one containing an infinite number of particles, the density of which is finite. A real macroscopic system is one that is sufficiently large that its equation of state is empirically indistinguishable from an ideal macroscopic system (Sewell 1986, p. 4).

assumption that the system's states are determinate in the sense that for any state description S, **S** makes either S or not-S true, then if T is an undecidable theory of **S**, some state of **S**, be it describable by S or by not-S will be underivable from T, hence unpredictable from T. A philosophical ancestor of this approach is C.D. Broad's position, although he wrapped the point around the concept of trans-ordinal laws: 'A trans-ordinal law would be a statement of the irreducible fact that an aggregate composed of aggregates of the next lower order in such and such proportions and arrangements has such and such characteristic and non-deducible properties.[4] (Broad 1925, pp. 77–78). Key terms here are 'irreducible' and 'non-deducible'.

A clear separation must be maintained between undecidability and computational irreducibility. The former concerns underivability in principle, the latter is about efficient prediction. Here is a definition of the latter concept restricted to cellular automata.

Definition (D): A cellular automaton computation of output O from input I that uses T timesteps is computationally irreducible just in case there is no way of computing O from I in fewer than T steps using another process that itself counts as a computation.

Here it is permissible to code O and I into representational forms that the alternative computational process requires.[5] Computational irreducibility is a finitistic concept, whereas undecidability has to have essentially non-finitistic aspects because of the Halting Problem.[6]

8.3 Results for Infinite Ising Lattices

In this section I provide the broad outline of the general method and technical results that are relevant to claims about emergence in this area. We can begin by considering a one-dimensional cellular automaton consisting in a linear array of cells, infinite in both directions. Each cell's state is a value of a binary variable, and at each time step the state of every cell is updated by a rule, the same for each cell, that is a function of the state of the cell and its two immediate neighbors. Updating is simultaneous for every cell. Here is one set of rules, for what is usually known as

---

[4] Brian McLaughlin asserts that Broad is using a semantic conception of deduction here; i.e. B is deducible from A if and only if when A is true, B is true (McLaughlin 1992), but I shall use the more standard proof-theoretic idea of deducibility.

[5] The concept of computational irreducibility can be found in Wolfram (1984). A coarse-graining approach to computational reducibility that is relevant to models in condensed matter physics can be found in Israeli and Goldenfeld (2006)

[6] The Halting Problem, which is actually a theorem, tells us that there is no effective procedure for determining in advance whether a program for a partial recursive function will halt on an arbitrary input or not.

the Rule 110 cellular automaton, where the arrow notation indicates that the middle cell of the triad is transformed to or maintained as a white cell in the top three cases and to a black cell in the bottom five cases.

■■■, ■□□ , □□□ → □

□□■, □■□, ■■□, □■■, ■□■ → ■

The rules for a given cellular automaton can alternatively be given in the form of an update function $f_u(x_t, y_t, z_t) = y_{t+1}$, where x, y, z are binary valued. It is known that a Rule 110 cellular automaton with a suitable input is universal; that is, with a properly chosen set of cell states along the top edge of the cellular automaton, the evolution of that cellular automaton under the rule set 110 will compute the value of any function that is considered computable under the standard Church-Turing thesis.[7] Now arrange each successive temporal stage of this one-dimensional cellular automaton in a time ordered sequence, with the initial t_0 state at the top edge. We then obtain a two-dimensional spatial array of cellular states that is unbounded at the bottom edge and that spatial array represents a computation with a specific input and (if the computation halts) a specific output. In philosophical jargon, the one dimensional A series of the cellular automaton has been represented as a two dimensional B series.

The basic idea is then to code the successive states of a cellular automaton onto ground states of a two dimensional periodic infinite Ising lattice so that the state of the lattice at a given time represents the entire temporal development of the chosen cellular automaton. Because there are inputs to the cellular automaton that represent the inputs to an undecidable problem there will be states of the corresponding Ising lattice that cannot be effectively predicted. A key result is Barahona (1982): For any cellular automaton rule R, there exists a periodic two dimensional Ising lattice the ground states of which represent the application of R for any input. Such a lattice consists in a regular grid of nodes at each of which a binary valued element is located. The lattice is an *Ising lattice* because it has a Hamiltonian given by:

$$H = \sum_{i,j} c_{ij} s_i s_j + \sum_i E_i s_i$$

where E_i represent values of an external field which may be highly non-uniform, the c_{ij} are coefficients representing interactions between the nodes, and $s_i \in \{-1, +1\}$.[8] The lattice is *periodic* because it is built from M x N blocks of nodes, each having the same interactions between nodes at corresponding places in the blocks, and the same interactions between the blocks. In the intended interpretation the nodes represent particles having spins and we restrict ourselves to nearest neighbour interactions between the spins, although that can be generalized, and our interest is

---

[7] See e.g. Weber (2012), pp. 65–66 for the Church-Turing thesis; Cook (2004) for the proof of the universality of Rule 110.

[8] For a detailed treatment of Ising models, see Binder and Heerman (1988), Humphreys (1994).

in the ground state—the state in which H is a minimum. For computational reasons, the values of the coefficients c_{ij} and E_i are restricted to rational values.[9]

These blocks are used to represent the update function for a specified cellular automaton. Omitting some technicalities, some of the nodes on the top edge of a block represent the argument values of the update function f_u and one of the nodes on the opposite edge represents the value of the update function for those inputs. By suitably arranging the blocks so that the outputs from some blocks are the inputs to others the entire temporal development of a cellular automaton can be represented by elements of the ground state of an appropriately chosen Ising lattice. Note that the Hamiltonian and hence the corresponding ground state depends on the set $\{E_i\}$ and so will usually differ depending upon the initial state of the cellular automaton.

It is important to stress that the values of the coefficients c_{ij} can include both positive and negative values. When a coefficient is negative, the spins tend to align so that the energy is lowered, whereas if the coefficient is positive, adjacent spins will tend to be in opposite directions. This representation is completely general. The familiar situation in which the system being modeled is a ferromagnet so that all of the c_{ij} are negative is a very special case of this more general representation. In the ferromagnetic case, the ground state occurs when all of the spins are aligned in the same direction and the solution of the ground state problem is of minimal complexity but in the general case, the ground state will be far more complex. This raises a specifically philosophical question. Physics has decided that the macroscopic properties corresponding to the extreme cases of ferromagnetism and antiferromagnetism count as genuine physical properties. What about the far more numerous intermediate cases? Do each of the macroscopic ground states of the lattice count as a novel physical property or is each of them simply a different value of a common property that includes the extreme cases? If the latter, then an inference from the unpredictability of a state to the conclusion that the state is emergent is unmotivated.

In Gu et al. (2009), the authors discuss a specific property of a cellular automaton, its 'prosperity'. The prosperity of a cellular automaton is the relative frequency of 1s among the cell states, averaged over all time steps from zero to infinity. That time averaged prosperity can be reproduced in the lattice by taking a spatial average over all the blocks that constitute the lattice and the spatial average can be identified with the overall magnetization M of the lattice.[10] Referring to our earlier definition, the prosperity is a macroscopic property of the entire lattice.

An important background result is then Rice's Theorem:

Let $A = \{n : \Phi_n^1 \in F\}$ where F is a set of at least one, but not all, partial recursive one-place functions and $\{\Phi_j^1 : j = 1, 2, \dots\}$ is an enumeration of all one-place partial recursive functions. Then the set A is not recursive.

---

[9] The lattices have no properties beyond those specified and so nothing specifically quantum mechanical should be imputed to them.

[10] The argument given in Gu et al. 2009 is not in the form of a formal proof and it omits details at various points. The exact way in which limits on the infinite lattice are justified is therefore hard to assess.

The intuition behind Rice's theorem is this: A is an index set that picks out a non-trivial property of partial recursive functions i.e. the property holds neither for all of them nor for none of them. Then there is no general algorithm that decides for any arbitrary partial recursive function whether or not its index is in A i.e. whether or not it has the property.

Here, for illustration, are two examples of the application of Rice's theorem. Let F be the set of all partial recursive functions that have finite domains. Then the problem—for an arbitrary Φ_j^1 determine whether it is a member of F or not a member of F i.e. decide whether or not its domain is finite—is recursively unsolvable. As a second example, let F be the set of all partial recursive functions that have a fixed point i.e. f $\in$ F if and only if there is some argument a such that f (a) = a. Then the problem of determining whether an arbitrary function Φ_j^1 is in F is recursively unsolvable.

Now consider the following decision problem: For an arbitrary input state to the Ising lattice, does M $\in$ [½, 1], where M is the overall magnetization? That problem must be undecidable because if not, the corresponding prosperity feature for the cellular automaton would be decidable, which it is not by Rice's Theorem. This undecidability result shows that the value of the macroscopic observable, here the overall magnetisation, cannot in all cases be deduced from the Ising model applied to the constituents of the system.[11]

8.4 Philosophical Consequences

We can now draw some philosophical conclusions. The formal results are about underivability in discrete models. Recalling our earlier distinction between inferential emergence and ontological emergence, the results at most bear on inferential emergence unless it is possible to argue that the models represent real physical systems. Some clarificatory remarks may be helpful here. First, I am not evaluating whether the claims of digital physics are correct. Digital physics claims that physical systems themselves, perhaps even the entire universe, are computational devices and that instead of scientific laws, we need to discover the algorithms driving the processes. Instead, we are dealing with discrete computational models without a metaphysical commitment to a digital universe.

Secondly, the often quoted remark by the statistician George Box that all models are wrong (Box 1976, p. 792) is incorrect and confuses incompleteness with falsity. Some discrete models do not involve idealizations or approximations and can be exact representations of their target systems. A counting model for steel ball bearings in a sealed container is exactly true and the fact that the model does not include the color of the ball bearings does not make it false, merely incomplete. In the case of emergence, exact truth is not required to establish the existence of

---

[11] This simplification of the argument omits details that can be found in Gu et al. (2009), p. 838.

ontological emergence on the basis of a model; it would be sufficient if we could argue that the Ising models represented, to within a satisfactory degree of approximation, the features needed for a physical system to display ontological emergence. As I shall show, the results we have discussed, and others similar to them, do not do that.[12] Thirdly, dealing with models rather than with real systems has one salient advantage, which is that we know exactly the rules that govern the system. We know that there is nothing more fundamental below the basic elements prescribed, and we know that nothing will affect the system other than what we have put in it. This third feature is important because, whereas Anderson was discussing the relation between fundamental physics, usually identified with the most recent version of high energy physics, and condensed matter physics, here we are discussing the relation between the fundamental rules governing the operation of an Ising lattice and a macroscopic property of that lattice. Because we are guaranteed that within the model there is no more fundamental level, this provides a parallel with Anderson's considerations, but it is as well to keep in mind that these models are not in any serious way describing fundamental physics. That said, the existence of in principle underivability anywhere in physics undermines the constructionist program.

One reason to doubt that these Ising models represent real systems is that the model involved in the undecidability result requires a two-dimensional spatially infinite lattice in order to represent the unbounded temporal development of the cellular automaton. It is therefore implausible, although it is not impossible, that it represents a real physical system.[13] Here we have a connection with Anderson's discussion of the importance of infinite limit results: 'It is only as the nucleus is considered to be a many-body system—in what is often called the $N \to \infty$ limit—that such behavior is rigorously definable...Starting with the fundamental laws and a computer, we would have to do two impossible things—solve a problem with infinitely many bodies, and then apply the result to a finite system—before we synthesized this behavior'. (Anderson op. cit. p. 395). In this regard, both Anderson's point and Gu et al.'s result fall into the same class of claims about derivability and the lack thereof as do limit results in models of condensed matter physics based on renormalization results, in the sense that they essentially rely on non-finite features. (See Batterman (2001), Butterfield (2011a, b)). The truth conditions for applying these models to finite physical systems must be given explicitly and these

---

[12] Gu et al. are clear both that their results, proved for two-dimensional lattices, can be extended to higher dimensions and that there are other interpretations of Ising models besides magnetic systems. Nevertheless, because appeals are made to features such as ground states and Hamiltonians, those other interpretations are irrelevant unless appropriate interpretations of H, E_i, and so on can be given that correspond to legitimate features of the concrete systems. I note that some well-known features of ferromagnets such as phase transitions play no role in the results here discussed.

[13] It is not impossible because if our universe contained at least two infinite dimensions, not necessarily spatial or temporal, within which the required features of the Ising lattice could be embedded, a real physical system would exist with undecidable properties. Finite dimensions with periodic boundary conditions would not do, but a two dimensional unbounded system that could be extended arbitrarily far in either dimension would.

formal results do not do that. They are proofs of how, in certain kinds of infinitely extrapolated worlds, unpredictable states can occur.

A second reason for doubting that these models represent reality is that the motivations for inferential emergence and for ontological emergence are very different. Whereas unpredictability in principle is the core of the inferentialist approach, ontological emergentists generally require that an emergent features be novel, and novelty has to be construed in an ontological fashion. Because the results in question are about unpredictable values of M and variations in the value of M from an extremal value of 0 or 1 to an intermediate value do not obviously produce ontological novelty, there is a question about whether the move from an undecidable value of M to an emergent feature is warranted. Gu et al. (2009, p. 835) had this to say: 'In 1972, P.W.Anderson...argu[ed] that complex systems may possess emergent properties difficult or impossible to deduce from a microscopic picture.... We show that for a large class of macroscopic observables, including many of physical interest, the value of those observables is formally undecidable...These results present analytic evidence for emergence'.

But new values of an existing property do not always yield new properties.[14] As noted earlier, ferromagnetism is a special case of overall magnetism that occurs when the spins in the sublattices are all aligned in the same direction. Antiferromagnetism occurs when adjacent spins are all aligned in opposite directions. Macroscopically, these are taken to be distinct physical properties, so the value of the overall magnetization can make a qualitative difference. We must therefore answer the question as to whether some other, far more complicated, patterns of spin alignments give rise to physically significant, novel macroscopic properties. Some are known, such as ferrimagnetism in which the nodes on sub-lattices are aligned in opposite directions but the magnet moments have different strengths, leading to an overall net magnetization. However, whether all microscopic patterns of spins give rise to physically important macroscopic properties is an issue that must be decided. That is, if we are interested in drawing conclusions about ontological emergence from results about underivability, we are faced with the problem of which microscopic patterns of spin correspond to real physical properties and which microscopic patterns of spin do not.[15] On a purely extensionalist account of properties, in which any set of spin values counts as a property, all of the spin arrangements create real but different macroscopic properties, but such purely extensionalist accounts seem too profligate and we appear to need additional criteria to identify the genuine physical properties. Unless we can provide such criteria, show that there is a process for determining whether a given microstate has that property, and provide some criterion of novelty for macroscopic states, we cannot move from the claim of inferential emergence to one of ontological emergence.

Next, if we do count new values of existing properties as novel, but continue to regard them as instances of an existing property, we shall have to abandon the

---

[14] For an early discussion of why novel values are not emergent, see Teller (1992).

[15] This is a real, rather than invented, example of the problem discussed in Dennett (1991).

common position that emergent features occur at a different level than non-emergent properties. Furthermore, as Margaret Morrison pointed out (personal communication) one reason that ferromagnetism is considered to be an emergent property is the presence of long range order associated with a phase transition, a feature that is often taken to be indicative of the transition to an emergent property. I note that antiferromagnetism is also accompanied by a (second order) phase transition which provides a principled reason for attributing emergent status to that state. Unless it can be shown that the undecidable values of the overall magnetization of the Ising model are associated with similar phase transitions, the undecidability accompanies a state lacking one of the central features of an emergent state in condensed matter physics. That is, we should have to show that there was some ontological correlate of undecidability that is part of the reason that the undecidable state is emergent. None of these challenges is insuperable, but they have to be answered before a claim of ontological emergence can be based on these results.

Nor can Anderson's position be used to support an inferentialist interpretation of emergence for these results, because being unable to compute certain *values* of an observable is not the situation with which Anderson was concerned. He was interested in situations within which a new descriptive apparatus, together with its own distinctive law statements, had to be used in order to represent and to predict states at the new level of complexity.[16] This kind of situation, in which the theoretical apparatus representing the microstates of the components of a system turns out to be inadequate to predict macroscopic states of that system is important enough to warrant further discussion in the next section.

8.5 The Axiomatic Method and Reduction

An ideal of long standing in the physical sciences is to represent the fundamental principles of a scientific field in axiomatic form and to take those principles as implicitly containing all of the knowledge in that field. If the first principles constitute a complete theory—and what 'complete' means in the context of scientific theories is a tangled issue—then the only role for empirical data is to provide the initial or boundary conditions for the system. Of course, testing and confirming the theory will require contact with data reports as well, but if the axioms are true, that aspect can be set aside. This ideal situation is unavailable in many cases and three consequences of its absence are of especial concern for us.

The first occurs when the values of certain parameters must be known in order for predictive use to be made of the theory. Borrowing a term from quantum

---

[16] The appeal to autonomous principles such as localization and continuous symmetry breaking seems to be a central feature of Robert Laughlin and David Pine's approach to emergence (Laughlin and Pine 2000).

chemistry, call a deductive or computational method *semi -empirical* if, in addition to the fundamental principles and parameters for the domain in question, some non-fundamental facts about the system must be estimated from empirical data rather than calculated from the fundamental principles and parameters. A simple example is when the value of the elasticity parameter in Hooke's Law cannot be estimated from first principles and must be known on the basis of measurement. Semi-empirical methods are often forced on us because of practical limitations but the results discussed in this paper show that, if these models, or ones like them, accurately represent systems in the physical world, then semi-empirical methods are unavoidable in areas beyond their origins in chemistry and this puts essential limits on constructivist knowledge..

The second consequence is that this situation places limits on the scope of theoretical knowledge and reveals an additional role for empirical input into scientific representations.[17] This consequence of undecidability has long been discussed in the philosophy of mathematics but it must be dealt with differently in science. The data needed to supplement the fundamental theory do not originate in intuition, as their correlates in mathematics are often supposed to do, but in measurement and other empirical procedures. This places them on a much less controversial epistemological basis. In order to know these macroscopic values we must either measure them directly or calculate them using a theory that represents the fundamental facts within a different conceptual framework.

The third consequence is that these results show limits on the hypothetico-deductive method. Suppose that we have a scientific theory T that is true of some states of a system but not of others and that all of the sentences that would falsify T fall into the class of sentences that are undecidable with respect to T. When we obtain empirical data and find that it is consistent with, for example, the truth conditions for a sentence ¬S but not with S, but neither S nor ¬S can be derived from T, that data cannot falsify T. S and ¬S are still falsifiable sentences in the standard sense that it is logically possible for each of them to be inconsistent with a sentence reporting the existence of empirically possible data but the theory T itself cannot be falsified using the hypothetico-deductive method together with modus tollens. Anderson's point is therefore not restricted to constructionist consequences but applies also to a particular variety of practical falsification.

Now let us return to Anderson's principal claim, that the truth of reduction does not entail the success of the constructionist project. I have assumed up to this point that the philosophical literature on reduction had not foreseen this point, but is that in fact true? It is correct that constructionist situations in which new concepts must be invented are not directly addressed by traditional theory reduction. This is

---

[17] By 'empirical' I do not mean 'empiricist' in any traditional sense of the term but the results of measurement, instrument output, or experiment having causal origins in the subject matter, however remote those are from human observational capacities. Discussions of such liberalized approaches can be found in Shapere (1982), Humphreys (1999), Bogen (2011).

because there is an implicit assumption in that tradition, regardless of whether we have an homogeneous or an inhomogeneous reduction in the sense of Nagel (1979), that the predicates employed at both the level to be reduced and the reducing level are already present as elements of the theories at those levels. In the simplest case the bridge laws will have the form $\forall x(J(x) \leftrightarrow I(x))$, where J and I are predicates occurring at the higher (reduced) and lower (reducing) levels respectively and both J and I exist before the bridge laws are constructed. Our reductive emphasis is usually on drilling downwards—for example, we already have the concept of water, we discover that at the molecular level we can perform a reduction, and with slight stipulative corrections we connect the existing higher level concept of water with the newer molecular concept of H_2O. Yet in some cases the higher level concepts do not yet exist and to represent a newly discovered phenomenon we must invent a new conceptual framework. Such situations are common in the social sciences and in astronomy. Smart mobs did not exist prior to the development of cheap and plentiful portable communications devices and such mobs have properties that distinguish them from traditional mobs, not the least of which is the ability for fast manouevering to outwit the police, to divide and reform at will, and for information to spread simultaneously rather than serially through its members. Whether the coming into existence of smart mobs is predictable or not, the core point is that we need to introduce a new concept to capture them at the macroscopic level. This a squarely a point about the representational power of a theory within which inferences are drawn—as I mentioned at the outset, I am interpreting Anderson's claims as being about our theories and models, and not directly about the world itself.

That said, it is not a part of the Nagelian tradition that the values of quantitative predicates at the higher level can be computed from values of quantitative predicates at the lower level. If those computations cannot be carried out, then the constructionist project fails because it requires that predictions and explananda at the higher level, including quantitative predictions and explananda, are derivable in practice. Definitions do not always provide algorithms that can be implemented; a simple example is the mean value of a stochastic quantity. Many such mean values can only be calculated numerically or by Monte Carlo methods and the definition must be supplemented by an algorithm for obtaining the relevant values using these methods.[18] Axiomatic formulations of theories tend to hide this fact, not just because they do not capture the details of specific applications but because many of them, such as axiomatic formulations of quantum theory and of probability are schema and the difficulties of specific applications only become apparent when the Hamiltonian or the probability distribution function replaces the placeholder function in the axioms.[19]

---

[18] For one such example, see Humphreys (1994).

[19] One exception to this neglect of the details of how axiomatic theories are applied is Suppes (1962), where a hierarchy of models is used to connect theory with data. Suppes explicitly allows that certain features of theory application do not lend themselves to formal treatment.

8.6 Finite Results

The arguments already discussed are undeniably interesting but their reliance on potentially infinite temporal processes and spatially infinite lattices makes them of mostly theoretical interest. What happens if we restrict ourselves to finite systems? Consider a common measure of computational difficulty.

Definition (M): A decision problem is NP (non-deterministic polynomial time) solvable just in case a proposed solution to the problem can be verified as a solution in polynomial time.[20] A decision problem NPC is NP-complete if every other NP solvable problem can be transformed into NPC in polynomial time.

Then results such as the following are available:

Computing the partition functions (and hence the exact energy levels) for finite sublattices of non-planar three dimensional and two dimensional Ising models are NP-complete problems. (Istrail 2000).

If, as is widely held, the class of NP problems cannot be identified with the class of P problems—those whose solutions can be computed in a time that is a poly-nomial function of the length of the input—then these NP problems pose severe difficulties in arriving at a solution in practice, plausibly the kind of situation with which Anderson was primarily concerend. Some caution is required about such results because computational complexity results concern worst case situations, so some instances of computing the energy levels can be easy, once again indicating that conclusions should not be drawn about a property type, but only about some of its values.[21]

8.7 Conclusions

The tendency in philosophy of science to place an emphasis on foundational studies and on what can be done in principle can lead to underestimating the power of Anderson's arguments for taking condensed matter physics and other non-funda-mental areas as autonomous and worthy of consideration in their own right. The technical results that have been established since his article appeared are undeniably impressive and important. Yet they leave unanswered a number of quintessentially philosophical questions that require a more detailed treatment. One can provide preliminary answers to some of those questions on the basis of the analysis given in this paper. The first is that one should not move from conclusions about the failure of constructivism and undecidability to conclusions about emergence without an explicit account of what counts as an entity being emergent and why. The failure of constructivism in a particular instance is not sufficient for emergence in the sense

---

[20] 'Time' here is to be taken as the number of computational steps required and the computation is referenced to any Turing machine-equivalent computational process.

[21] Further finite results can be found in Gu and Perales (2012).

that the inability in practice or in principle to compute values of a property is insufficient for the property itself to count as emergent. Secondly, the claim that there is a loss of understanding of these systems because of undecidability would only be true if understanding came through explanation and explanation required derivability. Yet the micro-processes that underlie the undecidable states of Ising lattices are similar for both decidable and undecidable values and those processes, whether causal or not, themselves provide at least partial understanding of how the macroscopic properties develop from the microscopic properties. Thirdly, these results support the position that the relationship between theory and data needs expanding so that there is a role for data in supplementing theoretical results in cases where it is impossible in principle or in practice to carry out the necessary derivations or computations. This suggests an expanded role for semi-empirical methods in physics and elsewhere.

Left as an open question is what counts as a novel physical property. It is widely held that a necessary condition for a property to count as emergent is that it be novel in an appropriate sense of 'novel'. If so, even the widespread inability to derive values of some macroscopic physical property does not by itself warrant a conclusion about the emergent status of that property. In addition, these considerations suggest a gap between the philosophical conception of reduction and emergence, which emphasizes properties, and the scientific conception which, at least within the cluster of literature discussed here, emphasizes values of those properties. This may well reflect disciplinary divisions of interest between the practice of philosophy, which is removed from the specifics of detailed derivations and computations, and the necessary emphasis of at least some areas of physics on the practicalities of application. Each side might gain from incorporating the concerns of the other.[22]

References

Anderson, P.W., Stein, D.L.: Broken symmetry, emergent properties, dissipative structures, life: are they related?. In: Anderson, P.W. (ed.) Basic Notions of Condensed Matter Physics, pp. 262–284. W. Benjamin, Menlo Park, CA (1984)

Anderson, P.W.: More is different: broken symmetry and the nature of the hierarchical structure of science. Science **177**, 393–396 (1972)

Barahona, F.: On the computational complexity of ising spin glass models. J. Phys. A: Math. Gen. **15**, 3241–3253 (1982)

Batterman, R: The Devil in the Details: Asymptotic Reasoning in Explanation, Reduction, and Emergence. Oxford University Press, New York (2001)

Beckermann, A., Flohr, H., Kim, J. (eds.): Emergence or Reduction?. Walter de Gruyter, Berlin (1992)

Binder, K., Heerman, D.: Monte Carlo Simulation in Statistical Physics. Springer, Berlin (1988)

Bogen, J: "Saving the Phenomena" and saving the phenomena. Synthese **182**, 7–22 (2011)

---

[22] Thanks to Jeremy Butterfield, Toby Cubitt, and Ashley Montanaro for comments about some of the formal results discussed here and to Margaret Morrison for comments that led to improvements in the paper as a whole. The conclusions drawn are entirely my own.

Box, G E.P.: Science and statistics. J. Am. Stat. Assoc. **71**, 791–799 (1976)

Broad, C.D.: Mind and Its Place in Nature. K.Paul, Trench and Trubner, London (1925)

Butterfield, J: Emergence reduction and supervenience: a varied landscape. Found. Phys. **41**, 920–959 (2011a)

Butterfield, J: Less is different: emergence and reduction reconciled. Found. Phys. **41**, 1065–1135 (2011b)

Cook, M.: Universality in elementary cellular automata. Complex Syst. **15**, 1–40 (2004)

Dennett, D.: Real patterns. J. Phil. **88**, 27–51 (1991)

Gu, M., Weedbrook, C., Perales, A., Nielsen, M.A.: More really is different. Phys. D **238**, 835–839 (2009)

Gu, M., Perales, A.: Encoding universal computation in the ground states of ising lattices. Phys. Rev. E **86**, 011116-1–011116-6 (2012)

Humphreys, P.: Extending ourselves. In: Massey, J. et al (ed.) Science at Century's End: Philosophical Questions on the Progress and Limits of Science. University of Pittsburgh Press, Pittsburgh (1999)

Humphreys, P.: How properties emerge. Phil. Sci. **64**(1997), 1–17 (1997)

Humphreys, P.: Numerical experimentation. In: Humphreys, P. (ed.) Patrick Suppes: Scientific Philosopher, vol 2, pp. 103–121. Kluwer Academic Press, Dordrecht (1994)

Israeli, N., Goldenfeld, N.: Coarse-graining of cellular automata, emergence, and the predictability of complex systems. Phys. Rev. E **73**, 026203-1–026203-17 (2006)

Istrail, S.: Statistical mechanics, three-dimensionality and NP-completeness: I. Universality of intractability for the partition functions of the ising model across Non-planar surfaces. In: Proceedings of the 32nd ACM Symposium on the Theory of Computing (STOC00), pp. 87–96. ACM Press, New York (2000)

Laughlin, R., Pines, D.: The theory of everything. Proc. Natl. Acad. Sci. **97**, 28–31 (2000)

McLaughlin, B.: The rise and fall of British Emergentism', pp. 49–93 in Beckerman et al. (1992)

Nagel, E.: Issues in the Logic of Reductive Explanations, Chapter 6 in Ernest Nagel, Teleology Revisited and Other Essays in the Philosophy and History of Science. Columbia University Press, New York (1979)

Sewell, G.L.: Quantum Theory of Collective Phenomena. The Clarendon Press, Oxford (1986)

Shapere, D.: The concept of observation in science and philosophy. Philos. Sci. **49**, 485–525 (1982)

Suppes, P.: Models of data. In: Nagel, E., Suppes, P., Tarski, A. (eds.) Logic, Methodology and Philosophy of Science: Proceedings of the 1960 International Congress, pp. 252–261. Stanford University Press, Stanford, CA (1962)

Teller, P.: A contemporary look at emergence, pp. 139–153 in Beckerman et al. (1992)

Weber, R.: Computability Theory. American Mathematical Society, Providence, Rhode Island (2012)

Wolfram, S.: Universality and complexity in cellular automata. Phys. D **10**, 1–35 (1984)

Chapter 9
Neither Weak, Nor Strong? Emergence and Functional Reduction

Sorin Bangu

9.1 Introduction

Past[1] and present[2] debates in scientifically informed metaphysics regard the concept of emergence as a chief concern. This paper addresses this topic from a novel perspective, aiming to contribute to its clarification by tackling it within the context of an examination of a special kind of idealization in physics. The overall goal of what follows is to signal the need for a refinement of a well-known distinction between two kinds of emergence, 'strong' and 'weak'. This distinction is central to virtually all ongoing debates on this topic, and for a good reason—it is a very natural one. Yet, I argue, this distinction is too coarse, and this shortcoming becomes visible when certain aspects of scientific theorizing (in particular in condensed matter physics) are taken into account. These aspects have been touched upon only occasionally in the vast literature on emergence, in part because, I suspect, the notion of emergence appropriate in this context has not yet been clearly delineated.

The paper is divided into two parts. The next section is mostly introductory, as it clarifies the main terms of the discussion in section III. I begin by laying out my understanding of what the distinction is; 'strong' and 'weak' emergence are terms of art, and such an elucidation is necessary as there is a bewildering variety of notions of emergence in the literature.[3] Generally speaking, I regard emergence in opposition to reduction.[4] Consequently, it is imperative to clarify what model of

---

[1] See especially Alexander (1920), Morgan (1923), Broad (1925), Pepper (1926).
[2] The collections edited by Beckermann *et al.* (1992), Gillet and Loewer (2001), Clayton and Davies (2006), and Humphreys and Bedau (2008) contain many influential recent articles.
[3] See O'Connor and Wong (2009) for an inventory.
[4] This is, again, a stipulation, as typically the notion is slightly broader than this.

S. Bangu (✉)
Department of Philosophy, University of Bergen, 7805, 5020 Bergen, Norway
e-mail: Sorin.Bangu@fof.uib.no

© Springer-Verlag Berlin Heidelberg 2015
B. Falkenburg and M. Morrison (eds.), *Why More Is Different*,
The Frontiers Collection, DOI 10.1007/978-3-662-43911-1_9

153

reduction I have in mind here, since, as is well known, there are a few such models on offer, including the Putnam-Kemeny-Oppenheim (1956, 1958) 'explanationist' model and the Nagel 'bridge law' model (1961). In what follows, however, I'll discuss a more recent (and, I believe, better articulated) model, due mainly to Kim (1998, 1999, 2006b), and called the 'functional' model (F-model hereafter).[5] As for the distinction strong v. weak emergence, although a number of authors made important contributions to establishing it (Humphreys 1997, 2006; McLaughlin 1992, 1997; Silberstein 1999; van Gulick 2001; Gillett 2002; O'Connor and Wong 2005; Bedau 1997; Bedau and Humphreys 2008, etc.), here I draw most explicitly on D. Chalmers' (2006) account of it. Against this backdrop, I shall end the first part of the paper by explaining how Kim's F-model construes strong emergence (in relation to weak emergence).

Next, I apply the F-model to a number of interesting and difficult cases from physics—yet involving very familiar phenomena, such as boiling and freezing. They are generically called 'phase transitions', or 'phase changes'. (There are two kinds of phase transitions, 'first-order' and 'continuous', but here I'm only concerned with the former). While a good deal of informative work on phase transitions has been published in the last 15 years (Humpherys 1997; Batterman 2002, 2005; Liu 1999, 2001; Ruetsche 2006; Callender 2001; Callender and Menon 2013; Bangu 2009; Butterfield 2011; Morrison 2012), none of these papers has tested the F-model on them yet, in order to show why, exactly, they look problematic for the F-reductionist.

It is the application of this reduction model to this class of phenomena that prompted the central thought I aim to articulate here, namely that the concepts of *weak* and *strong* emergence, as we find them in the contemporary metaphysics of science, are not able to accurately capture the subtlety of the conceptual situation found in the physics of phase change. A closer look at the details of these phenomena reveals that there is a clear sense in which they are emergent, and yet they are *neither weakly, nor strongly* emergent. In other words, their conceptual location is undecided on the current map of the metaphysical territory. As it will hopefully become clear later on, their status depends on how we understand the idealization relation between the theories describing the macro-level (classical thermodynamics) and the micro-level (statistical mechanics) of reality.

9.2 Types of Emergence and F-Reduction

As a first approximation, certain phenomena are taken to be emergent when they are irreducible, novel, unpredictable or unexplainable within the confines of a theoretical framework. Sometimes the concept of emergence is spelled out in

---

[5] D. Armstrong's and D. Lewis' writings in the 1960s and 1970s are of course among the first elaborations of the philosophical arguments grounding this model. See Armstrong (1968) and Lewis (1972).

mereological terms (very crudely, something—a whole—is emergent when it is, in *some* sense, more than the sum of its parts.) One might also say that there are aspects of the (emergent) whole which are completely different, new, and unexpected, given a complete knowledge of the nature of its parts. A more precise characterization of the emergence idea would be as follows (Humphreys 2006): an emergentist believes that the world contains things (typically: entities or properties) —such that (i) they are composed of (or accounted in terms of) other, more fundamental entities (or properties), and (ii) a complete reduction to these basic entities (or properties) is impossible.

Thus, on one hand, emergent phenomena are constituted by (generated from, or dependent on) underlying, lower-level processes and entities. (Some authors also talk in terms of the supervenience relation (e.g., Kim 2006a, b), but others refrain from doing so (e.g., van Gulick 2001), so this minimal characterization will leave this relation out for the moment.) Yet, on the other hand, emergent phenomena are, somehow, also autonomous from these underlying processes. Characterized in this way, it should be clear why emergence is a perennial philosophical puzzle: either the claim that there are emergent phenomena comes out as (logically?) inconsistent, or the existence of emergence looks like embodying the proverbial situation of getting something from nothing. (No wonder then that one of the most important contributions in this area is Kim's 1999 paper titled 'Making Sense of Emergence'!).

This puzzle is one of the important reasons for clarifying this elusive notion. But there is another reason, closely related to this difficulty. As Humphreys (2006, p. 191) further explains, we should care about emergence because, if this turns out to be a coherent and viable idea, it provides "direct evidence against the universal applicability of the generative atomism that has dominated Anglo-American philosophy in the last century", where "generative atomism" is "the view that there exist atomic entities, be they physical, linguistic, logical, or some other kind, and all else is composed of those atoms according to rules of combination and relations of determination." Hence, "the failure of various reductionist programs, especially that of physicalism, would be of significant interest to this program." Furthermore, it should also be noted that the existence of (strong) emergence would provide the grounds for defending the autonomy of the so-called 'special sciences' (chemistry, psychology, economics, etc.). As is well known, a good deal of the emergence literature pursues, in various forms, the project of showing that special sciences enjoy independence (in various forms) from the allegedly fundamental science of physics.

As I pointed out above, recent work on emergence distinguishes between 'strong' emergence (sometimes called 'ontological' emergence), and 'weak' emergence (or 'epistemological'). 'Strongly' emergent phenomena don't owe their emergent status to the limitations of our epistemic abilities. We call a higher-level phenomenon strongly emergent—and the precise formulation is: *emergent in relation to* a lower-level domain (Chalmers 2006)—when the higher-level phenomenon is ontologically dependent on the lower-level domain, and there are true statements about that phenomenon which are *impossible* to derive from the statements characterizing the entities and laws operating at the lower-level domain.

The contrast with 'weak' emergence becomes clearer now. The true statements about the higher-level weekly emergent phenomena are only *difficult* to derive, and thus often unexpected, given the principles governing the constitution and evolution of the lower-level domain. Importantly, the surprise, and novelty are ultimately grounded in our epistemic inability to predict (derive) the behavior of a certain aggregation of low-level components.

Returning to the strong type of emergence, recall that its hallmark was irreducibility in principle, that is, inability to predict regardless of how much computational power (human or non-human) we can mobilize. This kind of emergence seems intelligible, but then the important question is of course whether there are instantiations, or examples of such strongly emergent phenomena. Some authors dare to answer this question. Chalmers, for one, takes a resolute stance: "My own view is that the answer to this question is yes. I think there is exactly one clear case of a strongly emergent phenomenon, and that is the phenomenon of consciousness." (2006, 241) While not always so courageously voiced, this view is in fact widely spread. Many authors reflecting on this issue, including even some of those with strong physicalist sympathies (Kim, for one), believe that consciousness is the best candidate for an ultimately irreducible, strongly emergent phenomenon.

Now it is time to be more precise about what concept of reduction I will be presupposing in what follows. As I said above, the model of reduction at work here is due to Kim, who called it the 'functional' model (F-model), and designed it to remedy the flaws of the Nagel 'bridge law' model. One major advantage Kim claims for his F-model (see Kim 1998, 1999, 2006a, b), is the ability to avoid relying on bridge laws altogether—hence the model escapes the well-known problems plaguing Nagel's model.[6] Yet not everyone is convinced that this advantage is as advertised; some critics (e.g., Rueger 2000; Marras 2002, 2006; Fazekas 2009) are skeptic about the contention that the bridge laws are actually eliminated, and hence doubt that the gain over the Nagelian model is truly significant. In any case, because Kim's F-model has not been adequately tested on scientific examples, closer examination is required. After I explain the main idea behind it, I will stipulate that in what follows emergence just means the failure of reduction, where reduction is understood in the functional sense of Kim (hence I will call it 'F-reduction').

Here is the outline of the F-model. A higher level property P is reducible if and only if (step 1) P can be functionalized, i.e. defined in terms of its causal role, and (step 2) realizers of P can be found at a lower level, and there is a lower-level theory that explains (and permits calculations of) how the realizers operate.[7]

After introducing the model, Kim also illustrates how it works. One such illustration involves the biological, higher-level property 'being a gene'. Following the steps, we must first complete the functionalization step. Being a gene consists, functionally speaking, in "having some property (or being a mechanism) that performs a certain causal function, namely that of transmitting phenotypic

---

[6] For an inventory of these problems, see Sklar (1993).

[7] Kim breaks it down in three steps, but I collapsed the last two.

characteristics from parents to offsprings." (Kim 1999, p. 10). To accomplish step 2, we need to find the lower-level realizers of this property, as well as the lower-level theory that explains how they work. The realizers are of course the DNA molecules, while the theory is molecular biology. At this point, we know how to F-reduce phenotypic characteristics, genetically inherited features such as eye color, to a molecular basis.[8]

But Kim also takes up the hardest case, the reduction of qualia, such as pain. Here is how he applies the model. As expected, he points out that "...if pain is to be reduced to a brain process, the following must be accomplished: pain must first be given a functional definition and then we must identify its neural realizers...." (2006b, p. 553). In more detail, the two steps are now as follows: "The first step involves conceptual work: Is the concept of pain functionally definable and if so how should a functional definition of pain be formulated?" "The second step, that of discovering the realizers of pain, is up to empirical scientific research. It is in effect the research project of finding the neural correlates of conscious experiences..." And, as far as we know, the realizers of mental phenomena are brain phenomena; in particular, the realizers of pain will be the excitation of C-fibers.

It is important to note that, on Kim's view, the crucial step is the first one: "From a philosophical point of view, the crucial question, therefore, is whether pain can be given a functional characterization, in terms of physical input and behavioral output; the rest is up to science." (2006b, p. 553) This last point, that "the rest is up to science", reflects some widespread (philosophical) attitude that once the conceptual work is done, what's left is (more or less) tedious, mere computational labor. With an eye to the further development of the argument in this paper, it's good to remember that Kim is dismissing this second step in an F-reduction, since we will confront serious problems exactly at this step.

Now, as is well-known, Kim is doubtful that the first step can be completed for qualia, so this feature of consciousness can't be F-reduced. In his own words: "I am with those who do not believe pain and other sensory states ('qualia') can be given functional characterizations (e.g., Chalmers 1996)". (Kim 2006b, p. 553) For the purposes of what follows, however, the reasons he holds this view are irrelevant, so I won't elaborate on his skepticism. What interests me is that here we have a case in which what the model delivers (namely, emergence, or failure of F-reduction) fits many philosophers' intuitions—namely, that qualia are strongly emergent.

Before we move on to part III, let's take stock. Qualia, pain in particular, are emergent since they are not reducible in the F-model. And, importantly, *reduction fails at step* 1. Equally important, note that reduction seems achievable in the F-model once this first conceptual step is passed (and, for one thing, we do know what the realizers are). Therefore, recalling the classification of types of emergence, we can claim that, in this picture, *strong emergence = failure of F-reduction at step* 1.

---

[8] When presented with this argument, biologists and philosophers of biology alike are skeptical; but the topic of reduction in biology is a large and complex one, and it would certainly require a separate paper to discuss the success of the F-model in that context.

Once again, if step 1 is passed, all that's left is, at best, the possibility of weak emergence—that is, the essentially *epistemic* predicament encountered at step 2, involving the difficulties of computing the operations of the realizers within the lower-level theory.

9.3 Strong or Weak?

As I said at the outset, my argument will proceed by way of example. I will present a class of cases in which F-reduction is not blocked at step 1, but at step 2; and, moreover, it's not weak, but strong, in the sense that the difficulty faced is not merely computational (in other words, we fail to derive the higher-level phenomena from lower-level realizers *in principle*.)

First, let us ask to what extent does the F-model apply to cases beyond psychology and biology? Does it cover cases of potentially emergent properties encountered in physics? As it turns out, it does. Kim himself gives a hint of how we can understand the reduction of an interesting physical property, the transparency of water (Kim 1998, p. 100). According to the steps delineated above, we should first pin down the property to functionalize—and then functionalize it. The property is, clearly, 'being transparent', and the functionalization can be done in the following way: being transparent, for a certain substance or medium, is having "the *capacity* to transmit light rays intact" (Kim 1998, p. 100; italics added). What about step 2, the realizers and the theory? Kim doesn't address this explicitly, but the gap is relatively easy to fill. The lower-level realizers are of course photons, virtual photons, electrons, molecules of water, etc. And the theory that explains how these realizers work (how light interacts with matter) is quantum electrodynamics (QED).

One additional point worth-mentioning here is that this theory had to overcome serious difficulties before becoming canonical. Roughly speaking, the problem was that in its initial form QED made absurd predictions, such as that the charge of the electron is infinite. So the physicists who devised it (most notably Feynman, Tomonaga and Schwinger) had to invent a technique, called renormalization, to avoid these nonsensical consequences (See Feynman 1985). This problem is worth mentioning here because it reveals that difficulties in a functional reduction can *also* arise at step 2; and it is no accident that the cases I'm going to discuss in a moment encounter this type of difficulty too.

So, from now on the road ahead is as follows. I will apply the F-model to a class of physical properties called 'phase transitions'—comprising, as I said, everyday phenomena such as boiling and freezing. What I shall try to show is (i) that they are not reducible in the F-model, and (ii) this failure of reduction is different from the type of failure we encountered so far.

We are all familiar with boiling, in particular with boiling water. Is this property of water—that, in certain conditions, it boils—reducible according to the F-model? If we are to apply the model, the first issue we face is the functionalization of this property. One way to proceed is to notice that what happens when such a process

Fig. 9.1 A schematic phase diagram. At point O the substance is in all three phases simultaneously, hence its name, the 'triple' point. Beyond point C (the 'critical' point) the sharp distinction between the gaseous and liquid states vanishes

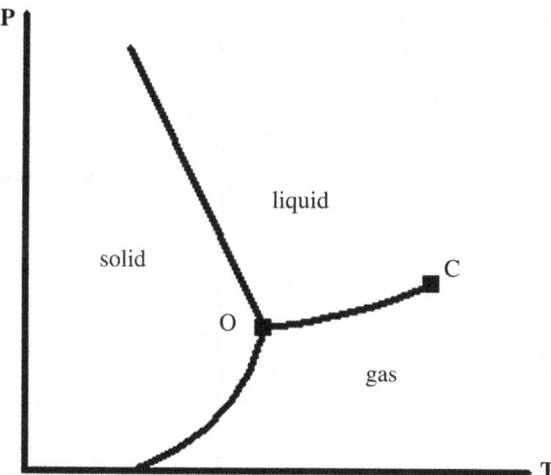

takes place is that water changes its aggregation state, from liquid to vapor. In other words, we deal with a qualitative change: the observable properties of the substance change significantly. From the higher-level perspective of thermodynamics, what is going on at the macro-level is that water crosses one of the lines in Fig. 9.1, called 'coexistence lines'.

The 'phase diagram' in Fig. 9.1shows the correlation between two of the thermodynamic, macro-level properties (temperature T and pressure P) and the state of aggregation of the substance (gas, solid and liquid). For any point in the diagram, corresponding to certain values of temperature and pressure, the substance is in a certain phase. If we keep the pressure constant while slowly increasing the temperature, water starting in liquid phase crosses the line OC and enters the vapor state. So, at the macro-level we can think of this property ('boiling') as consisting in having the capacity (to use Kim's word) to cross the coexistence line OC—from the liquid region toward the gas region. (To clarify, these coexistence lines consist of values for pressure and volume for which the substance is in two states simultaneously—or even three, at point O—and where just a slight variation in pressure or temperature takes the substance to the other state.)

Classical thermodynamics has the conceptual resources to make the above description more precise. (I follow Zemansky's 1968 classical textbook treatment.) We can define a quantity called the (Gibbs) 'free energy' of the system, represented by G, and equal to the difference

$$G = H - TS$$

where H, T and S are macro-level quantities (H is the enthalpy of the system, T is temperature, and S is the thermodynamical entropy). In these terms, we can say that crossing a coexistence line takes place when this quantity G (represented in Fig. 9.2 as dependent on T, at constant pressure) behaves in a special way.

Fig. 9.2 The Gibbs free
energy is plotted as a function
of temperature T. The graph
shows the phase transition as
a discontinuous change of the
first-order derivative. Drawn
after Zemansky (1968, p. 348)

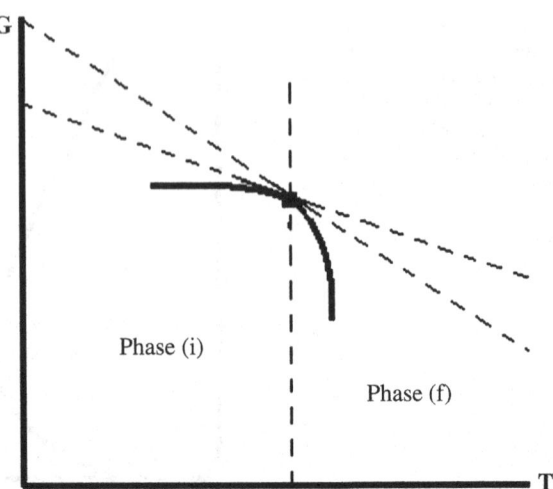

More precisely, the crossing takes place when this function G features a point at which the derivative (the tangent) varies discontinuously. In other words, the crossing (representing either boiling or freezing, it doesn't matter which for the present purposes) occurs when this function shows a 'kink', or 'sharp corner'. Such a point is called a 'singularity' ('non-analiticity'). So crossing, or changing phase at equilibrium, corresponds to a singularity (a non-analytic point) of function G, the Gibbs free energy. Now the requirement of step 1 in an F-reduction argument is satisfied. From the perspective of the higher-level theory (thermodynamics), and simplifying a bit, to boil a substance = to have its free energy feature a singular point.

This is what happens from the macro-level, or higher-level perspective, the perspective of Thermodynamics. Now we have to discuss step 2, and ask: what are the low-level realizers of this property and how do they work together to produce the phenomenon—that is, how does the low-level theory explain the realization of this property. There are no surprises awaiting for us here: the low-level, fundamental realizers are water molecules, while the theory must be Statistical Mechanics, more precisely (quantum) statistical mechanics. Hence, what we have to do is to show that a (immensely large) collection of such realizers is able to behave, in certain conditions, such that the system crosses a coexistence line.

Thus, the problem to solve at step 2 is clear: show that the statistical mechanical counterpart of the thermodynamical G features a singular point. But, as physicists know, this is prima facie impossible! David Ruelle signals the difficulty:

> So, here is a problem for the theoretical physicist: prove that as you raise or lower the temperature of water you have phase transitions to water vapor or to ice. Now, that's a tall order! We are far from having such a proof. In fact there is not a single type of atom or molecule for which we can mathematically prove that it will crystallize at low temperature. These problems are just too hard for us. (1991, pp. 123–124)

Very schematically, the situation is this (Liu 1999, 2001; Kadanoff 2000; Callender 2001; Batterman 2002; Bangu 2009). At the micro-level, that is, within statistical mechanics, we have to find a way to show that the function G has a singularity. But, when we express G in statistical mechanical terms we find that G depends on another quantity, Z, called 'the partition function' of the system (see Reif (1965, pp. 214–16) for details.) So, the new problem is to try to show that this quantity Z itself has a singular point. But Z is a sum of quantities specific for various states in which the micro-constituents might find themselves in. More precisely, the partition function can be written as

$$Z = \sum_r e^{-\beta E_r}$$

where E_r are the energies of the r (quantum) micro-states accessible to the system, $\beta = 1/kT$, and k is the Boltzmann constant. (We assume quantization, otherwise Z should be an integral).

The key-idea behind these details is straightforward: to show that the system undergoes a phase transition we have to show that G or, more precisely, the statistical mechanical expression of G, which depends on Z, has a singularity. Yet, crucially, as a matter of pure mathematics, G can't feature a singularity. The basis of this claim is a mathematical theorem. In essence, G depends on Z, and Z can't have singularities (non-analytic points) in so far as Z is a finite sum of analytic functions. It is because of this result that Ruelle called the problem "too hard".

We can thus conclude that phase transitions are not weakly emergent, since the derivation is impossible in principle—therefore they are a candidate for a strongly emergent phenomenon. This view is embraced by some notable physicists and philosophers. Lebowitz believes that phase transitions are "paradigms of emergent behavior" (1999, S346) and Prigogine makes the same point: "The existence of phase transitions shows that we have to be careful when we adopt a reductionist approach. Phase transitions correspond to emerging properties." (Prigogine 1997, p. 45). Physics popularization literature also supports the idea: "The emergence of solids, liquids and gases from a large collection of molecules, such as H_2O of water, ice, and steam, is an everyday example." (Close 2009, p. 113). Among philosophers, Liu calls them "truly emergent properties" (1999, S92).

What is important for my purposes here, however, is to note that they are *not* an instance of strong emergence due to the failure of F-reduction at step 1. Hence they are strongly emergent in a way that is not captured by the F-model (recall the point made above, that within the F-model strong emergence is identified with failures at step 1.)

Now I shall turn to the additional complications I mentioned above. Physicists were able, eventually, to get around the singularity problem, in a brilliant mathematical *tour de force* (Onsager 1944; Lee and Yang 1952, Yang and Lee 1952 are among the key contributions). They came up with a 'solution' (why scare quotes are needed will become apparent in a moment), and the way out was, in essence, to introduce an idealization that simplifies the original problem. They observed that considering a system composed of an infinite number of particles[9] (that is, working in the 'thermodynamic limit') allows them to find singularities in the statistical mechanical version of the function G (i.e., G(Z)). So it seemed that even step 2 in the F-reduction could be achieved after all.

We have now reached the puzzle that motivated the present paper in the first place. Suppose that, upon reflection on the ubiquitous phenomena of phase change, we would like to characterize this kind of property by using the conceptual vocabulary offered by the contemporary metaphysics of science. We thus ask, are they weakly emergent or strongly emergent? The problem, as I see it, is that no definite answer can be offered.

On one hand, if we are willing to make a certain assumption—that the system under study (say, the water in the kettle) is composed of an infinite number of molecules—then we should answer that they are not emergent at all, because step 2 is satisfied. The advantage of this position is that it does justice to our pre-scientific intuitions. Boiling and freezing don't seem something mysterious, and nobody ever assumed they were.

But, on the other hand, the previous assumption is problematic—an infinite system just doesn't exist in nature. So, since there is no way to rigorously explain (i.e., derive mathematically) how a phase transition occurs in a finite, real system[10], then these phenomena do appear to be strongly emergent.

The current solution via infinite idealization comes at a very high metaphysical price: a conflict with one of the central ontological assumptions of modern science, namely that the fundamental nature of reality is granulate and finite (in the sense that regardless of what kind of particles we divide chunks of matter into, the number of such particles is finite). In particular, we currently believe beyond any doubt that statistical mechanical systems are chunks of matter made of finitely many, smaller and stable, other chunks (molecules).

Hence, if we return to the key-question—what type of emergence do phase transitions instantiate, weak or strong—it should now be evident that neither answer is clearly right. It thus seems that we find ourselves in the proverbial is-it-a-bird?-is-it-a-plane? kind of situation, and we are confronting a new species of emergence, different from those already known.

We can get an even better grasp of what's going on in this case by making explicit what *would* count as a truly satisfactory solution to the non-analyticity problem (see also Bangu 2011). A genuine solution would be either (i) to find a way within statistical mechanics to derive the singularity without appealing to the infinite idealization, or (ii) to eliminate the singularity from the picture altogether.

---

[9] Further significant constraints are also imposed on this idealized model (called the Ising model), one of them being that the ratio between the number of particles in it and the volume it occupies is finite. When all these details are considered, it is not clear to what of the three types of idealizations identified by Weisberg (2007) this one belongs.

[10] As Kadanoff urges, "the existence of a phase transition requires an infinite system. No phase transitions occur in systems with a finite number of degrees of freedom." (2000, 238).

But both these options encounter serious difficulties. The second proposal would amount to changing a lot in the current physics, as it would involve theorizing phase transitions in a different way in statistical mechanics than in thermodynamics, thus objecting to the use of the thermodynamical definition in statistical mechanics.[11] The implementation of the first idea is not without problems either. It would also require a novel, different statistical mechanical treatment of singularities. Yet, unlike the other proposal, physicists find this one more palatable and in fact some notable efforts have recently been made in this direction (by Franzosi *et al.* 2000 among others), linking the singularities (of the micro-canonical entropy) to the thermodynamic phase transitions, the overall aim of this approach being to show that non-analyticities in the entropy somehow correspond to a change in the topology of configuration space.[12] Among philosophers, Callender and Menon 2013 are most optimistic about the outcome of these efforts and describe them as follows: "it is clear that the microcanonical ensemble does exhibit singularities even in the finite particle case and that there is a plausible research program attempting to understand phase transitions in terms of these singularities." (p. 217).

However, it is telling that they begin the paragraph in which this assessment is made by noting several "open questions" still to be answered by this approach, in particular "what topological criteria will be necessary and sufficient to define phase transitions, if any such criteria can be found." (2013, 217) And indeed, in a later paper Franzosi and Pettini (2004) (article not cited in Callender and Menon 2013) point out that the theorem they proved shows that

> ...a topology change (...) is a *necessary* condition for a phase transition to take place at the corresponding energy or temperature value (italics in original).

and that

> ...the converse of our Theorem is not true. There is not a one-to-one correspondence between phase transitions and topology changes; in fact, there are smooth, confining, and finite-range potentials (...) with even a very large number of critical points, and thus many changes in the topology (...) but with no phase transition. Therefore, an open problem is that of *sufficiency* conditions, that is, to determine which kinds of topology changes can entail the appearance of a [phase transition] (italics in original).

---

[11] As Callender memorably suggested, we should not 'take thermodynamics too seriously': "After all, the fact that thermodynamics treats phase transitions as singularities does not imply that statistical mechanics must too." (2001, 550).

[12] Slightly more precisely, Franzosi *et al.* (2000, p. 2774) describe their central idea as follows: "a major topology change (...) is at the origin of the phase transition in the model considered." Furthermore: "suitable topology changes of equipotential sub-manifolds of configuration space can entail thermodynamic phase transitions(...). The method we use, though applied here to a particular model, is of general validity and it is of prospective interest to the study of phase transitions in those systems that challenge the conventional approaches, as might be the case of finite systems." Kastner (2008) is another paper discussing this issue, also mentioned by Callender and Menon (2013).

It is reservations like these which confirm the impression that Callender and Menon's optimism—that "statistical mechanics might well have the resources to adequately represent these discontinuities without having to advert to the thermodynamic limit" (2013, p. 217)—is currently a little more than an article of faith.[13]

9.4 Conclusion

By insisting on a mathematically rigorous matching of thermodynamics and statistical mechanics, and on the finiteness of statistical mechanical systems, one can declare phase transitions strongly emergent phenomena—since mathematical singularities can't be derived. But one can also retort that the singularities *are* derived after all (in an infinite system), and thus these phenomena are not strongly emergent, but rather weakly emergent. This dilemma suggests that this case is actually very special, and not covered by the conceptual lexicon of contemporary metaphysics: it is neither a clear example of strong emergence, nor a clear case of weak emergence, and surely not an unproblematic success of F-reduction. Thus, the lesson thermal physics teaches the metaphysicians here is that they don't (yet?) possess a suitable notion of emergence to characterize these situations.

Acknowledgments I presented versions of this paper in several places (Urbana IL, Peoria IL, Boston MA, Fort-Wayne IN, and Berlin), and my audiences have been extremely helpful in commenting on, and criticizing previous versions of the paper. I thank a group of mathematical physicists at Bradley University (and my host there, Vlad Niculescu), Rob Cummins, Laura Ruetsche, Jessica Wilson, Peter Bokulich, Alisa Bokulich, John Stachel, Iulian Toader, Ioan Muntean, Mark Zelczer, Alex Manafu, Paul Humphreys, Craig Callender, Bob Batterman, Margie Morrison and Brigitte Falkenburg. I am of course responsible for all remaining philosophical or scientific errors.

References

Alexander, S.: Space, Time, and Deity, vol. 2. Macmillan, London (1920)
Armstrong, D.M.: A Materialist Theory of the Mind. Humanities Press, New York (1968)
Bangu, S.: Understanding thermodynamic singularities. Phase transitions, data and phenomena. Philos. Sci. **76**(4), 488–505 (2009)

---

[13] The topology change is not the only approach Callender and Menon discuss in their (2013). Two others, the 'back-bending' in the microcanonical caloric curve (Sect. 3.1.1) and the perpendicular distribution of zeros (Sect. 3.1.2) are also mentioned, but these approaches too fall under the same disclaimer, that "Probably none of the definitions provide necessary and sufficient conditions for a phase transition that overlaps perfectly with thermodynamic phase transitions", while also adding, without really explaining the claim, that "That, however, is okay, for thermodynamics itself does not neatly characterize all the ways in which macrostates can change in an 'abrupt' way" (2013, 210).

Bangu, S.: On the Role of Bridge Laws in Intertheoretic Relations. Philos. Sci. **78**(5), 1108–1119 (2011)

Batterman, R.: The Devil in the Details: Asymptotic Reasoning in Explanation, Reduction, and Emergence. Oxford University Press, Oxford (2002)

Batterman, R.: Critical phenomena and breaking drops: infinite idealizations in physics. Stud. Hist. Philos. Mod. Phys. **36**, 225–244 (2005)

Beckermann, A., Flohr, H., Kim, J. (eds.): Emergence or Reduction?. Walter de Gruyter, Berlin (1992)

Bedau, M.: Weak Emergence Philosophical Perspectives, 11: Mind, Causation, and World, pp. 375–399. Blackwell, Oxford, (1997)

Bedau, M.: Downward Causation and Autonomy in Weak Emergence, pp. 155–188. Bedau and Humphries (2008)

Bedau, M.A., Humphreys, P. (eds.): Emergence: Contemporary Readings in Philosophy and Science. MIT Press, Cambridge (2008)

Broad, C.D.: The Mind and Its Place in Nature, 1st (edn.) Routledge and Kegan Paul, London (1925)

Butterfield, J.: Less is different: emergence and reduction reconciled. Found. Phys. **41**, 1065–1135 (2011)

Callender, C.: Taking thermodynamics too seriously. Stud. Hist. Philos. Mod. Phys. **32**, 539–553 (2001)

Callender, C., Menon T.: Turn and Face the Strange … Ch-ch-changes: Philosophical Questions Raised by Phase Transitions. In: Batterman, R.W. (ed.) Oxford Handbook for the Philosophy of Physics. Oxford Univ. Press, New York (2013)

Chalmers, D.: The Conscious Mind. In Search of a Theory of Conscious Experience. Oxford University Press, New York (1996)

Chalmers, D.: Strong and Weak Emergence in Clayton and Davies (2006)

Clayton, P., Davies, P. (eds.): The Re-Emergence of Emergence. Oxford University Press, Oxford (2006)

Close, F.: Nothing. A Very Short Introduction. Oxford Univ. Press, Oxford (2009)

Fazekas, P.: Reconsidering the role of bridge laws in inter-theoretical reductions. Erkenntnis **71**(3), 303–322 (2009)

Feynman, R.: QED: The Strange Theory of Light and Matter. Princeton University Press, Princeton (1985)

Franzosi, R., Pettini, M., Spinelli, L.: Topology and phase transitions: paradigmatic evidence. Phys. Rev. Lett. **84**, 2774–2777 (2000)

Franzosi, R., Pettini, M.: Theorem on the origin of phase transitions. Phys. Rev. Lett. **92**, 060601 (2004) Available:http://journals.aps.org/prl/pdf/10.1103/PhysRevLett.92.060601

Gillet, C., Loewer B. (eds.): Physicalism and Its Discontents. Cambridge Univ. Press, Cambridge (2001)

Gillet, C.: The varieties of emergence: their purposes, obligations and importance. Grazer Philosophische Studien **65**, 89–115 (2002)

Humphreys, P.: How properties emerge. Philos. Sci. **64**, 1–17 (1997)

Humphreys, P.: Emergence pp. 190–194 In: Borchert, D. (ed.) The Encyclopedia of Philosophy (2nd edn., vol. 3). MacMillan, New York (2006)

Kadanoff, L.: Statistical Physics. World Scientific, Singapore (2000)

Kastner, M.: Phase transitions and configuration space topology. Rev. Mod. Phys. **80**, 167–187 (2008)

Kemeny, J.G., Oppenheim, P.: On reduction. Philos. Stud. **7**, 6–18 (1956)

Kim, J.: Mind in a Physical World. MIT Press, Cambridge (1998)

Kim, J.: Making sense of emergence. Philos. Stud. **95**, 3–36 (1999)

Kim, J. (2006a). 'Being Realistic about Emergence' in Clayton and Davies (2006)

Kim, J.: Emergence: core ideas and issues. Synthese **151**(3), 347–354 (2006b)

Lebowitz, J.L.: Statistical mechanics: a selective review of two central issues. Rev. Mod. Phys. **71**, S346–S347 (1999)

Lee, T.D., Yang, C.N.: Statistical theory of equations of state and phase transitions. II.Lattice gas and Ising model. Phys. Rev. **87**, 410–419 (1952)

Lewis, D.: Psychophysical and theoretical identifications. Australas. J. Philos. **50**(3), 249–258 (1972)

Liu, C.: Explaining the emergence of cooperative phenomena. Philos.Sci. PSA 1998 **66**, S92–S106 (1999)

Liu, C.: Infinite systems in SM explanations: thermodynamic limit, renormalization (semi-) groups, and irreversibility. Philos. Sci. PSA 2000 **68**, S325–S344 (2001)

Marras, A.: Kim on reduction. Erkenntnis **57**(2), 231–257 (2002)

Marras, A.: Emergence and reduction: reply to kim. Synthese **151**(3), 561–569 (2006)

McLaughlin, B.: The rise and fall of British emergentism. In: Beckermann et al. (ed) Emergence or Reduction? (1992)

McLaughlin, B.: Emergence and supervenience. Intellectica **2**, 25–43 (1997)

Morgan, C.L.: Emergent Evolution. Williams and Norgate, London (1923)

Morrison, M.: Emergent physics and micro-ontology. Philos. Sci. **79**, 141–166 (2012)

Nagel, E.: The Structure of Science. Harcourt, Brace and World, New York (1961)

O'Connor, T., Wong, H.: The metaphysics of emergence. Noûs **39**, 658–678 (2005)

O'Connor, T., Wong, H.: Emergent Properties. In: Zalta, E.N. (ed.) The Stanford Encyclopedia of Philosophy (Spring 2009 Edition). http://plato.stanford.edu/archives/spr2009/entries/properties-emergent/ (1992)

Onsager, L.: Crystal statistics. I. A two-dimensional model with an order-disorder transition. Phys. Rev. ser. 2, **65**, 117–149 (1944)

Pepper, S.: Emergence. J. Philos. **23**, 241–245 (1926)

Prigogine, I.: End of Certainty. The Free Press, New York (1997)

Putnam, H., Oppenheim, P.: The unity of science as a working hypothesis'. Minn. Stud. Philos. Sci. **2**, 3–36 (1958)

Reif, F.: Statistical and thermal physics. McGraw-Hill, New York (1965)

Rueger, A.: Robust supervenience and emergence. Philos. Sci. **67**, 466–489 (2000)

Ruelle, D.: Chance and Chaos. Princeton Univ. Press, Princeton (1991)

Ruetsche, L.: Johnny's so long at the ferromagnet. Philos. Sci. **73**, 473–486 (2006)

Silberstein, M., McGeever, J.: The search for ontological emergence. Philos. Quart. **49**, 182–200 (1999)

Sklar, L.: Physics and Chance. Cambridge University Press, Cambridge (1993)

Van Gulick, R.: Reduction, emergence and other recent options on the mind/body problem: a philosophic overview. J. Conscious. Stud. **8**(9–10), 1–34 (2001)

Weisberg, M.: Three Kinds of Idealization. J. Philos. **104**(12), 639–659 (2007)

Yang, C.N., Lee, T.D.: Statistical theory of equations of state and phase transitions. I. Theory of condensation. Phys. Rev. **87**, 404–409 (1952)

Zemansky, M.W.: Heat and Thermodynamics: an Intermediate Textbook (5th edn.) McGraw-Hill, New York (1968)

Part III
Parts and Wholes

Chapter 10
Stability, Emergence and Part-Whole Reduction

Andreas Hüttemann, Reimer Kühn and Orestis Terzidis

10.1 Introduction

Often we can describe the macroscopic behaviour of systems without knowing much about the nature of the constituents of the systems let alone the states the constituents are in. Thus, we can describe the behaviour of real or ideal gases without knowing the exact velocities or places of the constituents. It suffices to know certain macroscopic quantities in order to determine other macroscopic quantities. Furthermore, the macroscopic regularities are often quite simple. Macroscopic quantities are often determined by only a few other macroscopic quantities. This fact is quite remarkable, as Jerry Fodor noted:

> Damn near everything we know about the world suggests that unimaginably complicated to-ings and fro-ings of bits and pieces at the extreme micro-level manage somehow to converge on stable macro-level properties. [...] [T]he 'somehow', really is entirely mysterious [...] why there should be (how there could be) macro level regularities at all in a world where, by common consent, macro level stabilities have to supervene on a buzzing, blooming confusion of micro level interactions (Fodor 1997, p. 161).

The puzzle is that on the one hand macroscopic behaviour supervenes on the behaviour of the constituents, i.e. there is no change in the macroscopic behaviour without some change at the micro-level. On the other hand not every change in the

A. Hüttemann (✉)
Philosophisches Seminar, Albertus Magnus Platz, 50923 Cologne, Germany
e-mail: ahuettem@uni-koeln.de

R. Kühn
Department of Mathematics, King's College London, Strand, London WC2R 2LS, UK

O. Terzidis
Technologie-Management und Innovation, Institut für Entrepreneurship, Karlsruhe Institute of Technology, Fritz-Erler-Str. 1-3, 76133 Karlsruhe, Germany

© Springer-Verlag Berlin Heidelberg 2015
B. Falkenburg and M. Morrison (eds.), *Why More Is Different*,
The Frontiers Collection, DOI 10.1007/978-3-662-43911-1_10

states of the constituents leads to a change on the macro-level. To some extent the macro-behaviour is independent of what is going on at the micro-level. The questions we will address in this paper is whether there is an explanation for the fact that as Fodor put it the micro-level "converges on stable macro-level properties", and whether there are lessons from this explanation for other issues in the vicinity.

Various metaphors have been used to describe this peculiar relation of the micro-level to the macro-level. The macro-behaviour has been said to be "autonomous" (Fodor 1997), it has been characterized as "insensitive to microscopics" or due to "higher organizing principles in nature" (Laughlin and Pines 2000, p. 261) and above all macroscopic behaviour is said to be "emergent", where emergence implies novelty, unexplainability and irreducibility. There is no consensus about how best to define the concept of emergence—partly due to the fact that there is no uncontroversial set of paradigmatic examples and partly due to the fact the concept of emergence plays a role in a variety of philosophical and scientific contexts (For a discussion see Humphreys and Bedau 2008, Introduction).

As a consequence of these conceptual difficulties we will in the first part of the paper focus on the notion of the *stability* of macro-phenomena. One obvious advantage of this terminological choice consists in the fact that stability as opposed to emergence allows for degrees. Even though there is no precise concept of emergence it seems uncontroversial that whether or not some behaviour falls under this concept should be an all or nothing affair. The very word "stability" by contrast allows for behaviour to be more or less stable. However, in the second part of the paper we will also take up some issues that play a role in debates about emergence.

Even though the concept of stability has been used in various different senses within the philosophy of science recently, there is a clear conceptual core: Stability is a property we attribute to entities (things, properties, behaviour, sets, laws etc.) if the entity in question does not change even though other entities, that are specified, do change. This definition captures the notions of stability that have been recently been discussed in various contexts. Woodward and Mitchell call laws or causal generalizations "stable" relative to background conditions if they continue to hold, even though the background conditions change (Mitchell 2003, p. 140; Woodward 2007, p. 77). Lange calls a set of statements stable if their truth-value would remain the same under certain counterfactual changes (Lange 2009, p. 29). What we are interested in this paper is the stability of the behaviour of macro-systems vis-à-vis changes with respect to the micro-structure of the system. So the changes we consider concern not external or background conditions but rather system-internal features.

An explanation for the stability of macro-phenomena is not only an interesting project in itself. We do think that our discussion of stability throws some light on some issues that play a role in debates about emergence, as we will explain from Sect. 10.5 onwards. Furthermore our account of stability is of wider significance for other areas of philosophy. It is apparent that Fodor's seminal papers on the autonomy of the special sciences is largely motivated by the observation that macro-behaviour is stable under some changes on the micro-level. This in turn motivated the position of 'non-reductive' physicalism in the philosophy of mind.

Even though what we will finally present is not a non-reductive but rather a reductive account of stability, this account nevertheless vindicates what we take to be two of the core-intuitions that motivate non-reductive physicalism: (1) There are macro-level properties that are distinct from micro-level properties. (2) Macro-level properties are dependent on micro-level properties (See Rudder-Baker 2009, p. 110 for a similar characterisation of non-reductive physicalism that does not refer explicitly to the failure of reduction). Even though we will not discuss this point in any detail we would like to suggest that it is the stability of macro-behaviour that makes distinct macro-properties possible.

Fodor's question, as to how it is possible that extremely complex behaviour at an atomistic level could (for many macroscopic systems we know of) give rise to stable macroscopic behaviour has in recent years been transformed into a wider and more ambitious question, viz. how it is possible that even microscopically very different systems manage to exhibit the same type of stable macroscopic behaviour. In a still wider context it has been discussed by Batterman, most notably in his paper "Multi-realizablity and Universality" (2000) as well as Batterman (2002). His central claim is that the notion of universality used to describe the surprising degree of insensitivity of critical phenomena in the vicinity of continuous phase transition points can also be used to explain how special science properties can be realized by quite heterogeneous systems. Indeed, the renormalization group (RNG) explanation of the universality of critical phenomena was advocated as a paradigm that could account for the multi- realizability of macro-behaviour in general. While we agree that RNG provides an explanation of a certain kind of stability of macro-behaviour, we wish to point out that the case of critical phenomena is very special, in that it is restricted to phenomena that obtain in the vicinity of 2nd order phase transitions, but not elsewhere. Batterman himself is aware of this problem (See Batterman 2011). A similar remark pertains to Morrison's recent paper on emergence (Morrison 2012). Her account of emergence and stability in terms of symmetry breaking pertains only to a restricted class of those systems that exhibit stable behaviour.

For a more general explanation of the stability of macro-phenomena of individual systems, but also across classes of similar systems, other principles must therefore be invoked. In this paper we propose to rationalise the stability of macro-behaviour by pointing out that observations of macro-behaviour are usually observations on anthropomorphic scales, meaning that they are results of coarse-grained observations in both space and time. That is, they involve averages, both over individual behaviours of huge numbers of atomistic components constituting the system under study, and averages of their behaviour over time scales, which are very large compared to characteristic time-scales associated with motion at the micro-level. In fact we shall point out that large numbers of constituent atomistic components, beyond their role in observations on anthropomorphic scales, are also a prerequisite for the very existence and stability of ordered states of matter, such as crystals or magnetic materials.

Just to give an impression of orders of magnitude involved in descriptions of macroscopic amounts of matter, consider a cubic-millimetre of a gas at ambient temperature and pressure. It already contains approximately 2.7×10^{16} gas molecules, and the distance between constituents particles is so small that the typical time between two successive scattering events with other molecules would for each of the molecules be of the order of 10^{-10} s, entailing that equilibration times in such systems are very short.

Indeed, the traditional way of taking the time-averaging aspect associated with observations of macro-behaviour into account has been to consider results of such observations to be properly captured by values characteristic of thermodynamic equilibrium, or—taking a microscopic point of view—by equilibrium statistical mechanics. It is this microscopic point of view, which holds that a probabilistic description of macroscopic systems using methods of Boltzmann-Gibbs statistical mechanics is essentially correct that is going to form a cornerstone of our reasoning. Indeed, within a description of macroscopic systems in terms of equilibrium statistical mechanics it is essential that systems consist of a vast number of constituents in order to exhibit stable, non-fluctuating macroscopic properties and to react in predictable ways to external forces and fields. In order for stable ordered states of matter such as crystals, magnetic materials, or super-conductors to exist, numbers must (in a sense we will specify below) in fact be so large as to be "indistinguishable from infinity".

Renormalization group ideas will still feature prominently in our reasoning, though somewhat unusually in this context with an emphasis on the description of behaviour away from criticality.

We will begin by briefly illustrating our main points with a small simulation study of a magnetic system. The simulation is meant to serve as a reminder of the fact that an increase of the system size leads to reduced fluctuations in macroscopic properties, and thus exhibits a clear trend towards increasing stability of macroscopic (magnetic) order and—as a consequence—the appearance of ergodicity breaking, i.e. the absence of transitions between phases with distinct macroscopic properties in finite time (Sect. 10.2). We then go on to describe the mathematical foundation of the observed regularities in the form of limit theorems of mathematical statistics for independent variables, using a line of reasoning originally due to Jona-Lasinio (1975), which relates limit theorems with key features of large-scale descriptions of these systems (Sect. 10.3). Generalizing to coarse-grained descriptions of systems of interacting particle systems, we are lead to consider RNG ideas of the form used in statistical mechanics to analyse critical phenomena. However, for the purpose of the present discussion we shall be mainly interested in conclusions the RNG approach allows to draw about system behaviour away from criticality (Sect. 10.4). We will briefly discuss the issue how the finite size of actual systems affects our argument (Sect. 10.5). We will furthermore discuss to what extent an explanation of stability is a reductive explanation (Sect. 10.6) and will finally draw attention to some interesting conclusions about the very project of modelling condensed matter systems (Sect. 10.7).

10.2 Evidence from Simulation: Large Numbers and Stability

Let us then begin with a brief look at evidence from a simulation, suggesting that macroscopic systems—which according to our premise are adequately characterised as stochastic systems—need to contain large numbers of constituent particles to exhibit stable macroscopic properties.

This requirement, here formulated in colloquial terms, has a precise meaning in the context of a description in terms of Boltzmann-Gibbs statistical mechanics. Finite systems, are according to such a description, ergodic. They would therefore attain all possible micro-states with probabilities given by their Boltzmann-Gibbs equilibrium distribution and would therefore in general *also* exhibit fluctuating *macroscopic* properties as long as the number of constituent particles remains finite. Moreover ordered phases of matter, such as phases with non-zero magnetization or phases with crystalline order would not be absolutely stable, if system sizes were finite: for finite systems, ordered phases will always also have a finite life-time (entailing that ordered phases in finite systems are *not* stable—in a strict sense. However, life-times of ordered states of matter *can* diverge in the limit of infinite system size.

Although real world systems are clearly *finite*, the numbers of constituent particles they contain are surely unimaginably large (recall numbers for a small volume of gas mentioned above), and it is the fact that they are so very large which is responsible for the fact that fluctuations of their macroscopic properties are virtually undetectable. Moreover, in the case of systems with coexisting phases showing different forms of macroscopic order (such as crystals or magnetic systems with different possible orientations of their macroscopic magnetisation), large numbers of constituent particles are also responsible for the fact that transitions between different manifestations of macroscopic order are sufficiently rare to ensure the stability of the various different forms of order.

We are going to illustrate what we have just described using a small simulation of a magnetic model-system. The system we shall be looking at is a so-called Ising ferro-magnet on a 3D cubic lattice. Macroscopic magnetic properties in such a system appear as average over microscopic magnetic moments attached to 'elementary magnets' called spins, each of them capable of two opposite orientations in space. These orientations can be thought of as parallel or anti-parallel to one of the crystalline axes. Model-systems of this kind have been demonstrated to capture magnetic properties of certain anisotropic crystals extremely well.

Denoting by $s_i(t) = \pm 1$ the two possible states of the i-th spins at time t, and by $s(t)$ the configuration of *all* $s_i(t)$, one finds the macroscopic magnetisation of a system consisting of N such spins to be given by the *average*

$$S_N(s(t)) = \frac{1}{N} \sum_{i=1}^{N} s_i(t). \tag{10.1}$$

In the model system considered here, a stochastic dynamics at the microscopic level is realised via a probabilistic time-evolution (Glauber Dynamics), which is known to converge to thermodynamic equilibrium, described by a Gibbs-Boltzmann equilibrium distribution of micro-states

$$P(s) = \frac{1}{Z_N} \exp[-\beta H_N(s)] \tag{10.2}$$

corresponding to the "energy function"

$$H_N(s) = -\sum_{(i,j)} J_{ij} s_i s_j. \tag{10.3}$$

The double sum in this expression is over all possible pairs (i, j) of spins. In general, one expects the coupling strengths J_{ij} to decrease as a function of the distance between spins i and j, and that they will tend to be negligibly small (possibly zero) at distances larger than a maximum range of the magnetic interaction. Here we assume that interactions are non-zero only for spins on neighbouring lattice sites. One can easily convince oneself that positive interaction constants, $J_{ij} > 0$, encourage parallel orientation of spins, i.e. tendency to macroscopic ferromagnetic order. In (10.2) the parameter β is a measure of the degree of stochasticity of the microscopic dynamics, and is inversely proportional to the absolute temperature T of the system; units can be chosen such that $\beta = 1/T$.

Figure 10.1 shows the magnetisation (10.1) as a function of time for various small systems. For the purposes of this figure, the magnetisation shown is already averaged over a time unit.[1]

The first panel of the figure demonstrates that a system consisting of $N = 3^3 = 27$ spins does not exhibit a stable value of its magnetisation. Increasing the number of spins $N = 4^3 = 64$ that the system appears to prefer values of its magnetisation around $S_N(s(t)) \simeq \pm 0.83$. However, transitions between opposite orientations of the magnetization are still very rapid. Increasing numbers of spins further to $N = 5^3 = 125$, one notes that transitions between states of opposite magnetization become rarer, and that fluctuations close to the two values $S_N(s(t)) \simeq \pm 0.83$ decrease with increasing system size. Only one transition is observed for the larger system during the first 10^4 time-steps. However, following the time-evolution of the magnetisation of the $N = 5^3 = 125$ system over a larger time span, which in real-time would correspond to approximately 10^{-7} s, as shown in the bottom right panel, one still observes several transitions between the two orientations of the magnetization, on average about 10^{-8} s apart.

---

[1] The time unit in these simulations is given by the time span during which every single spin has on average once been selected for an update of its state. It is this unit of time which is comparable for systems of different sizes (Binder and Stauffer 1987, pp. 1–36); it would correspond to a time-span of approximately 10^{-12} s in conventional units.

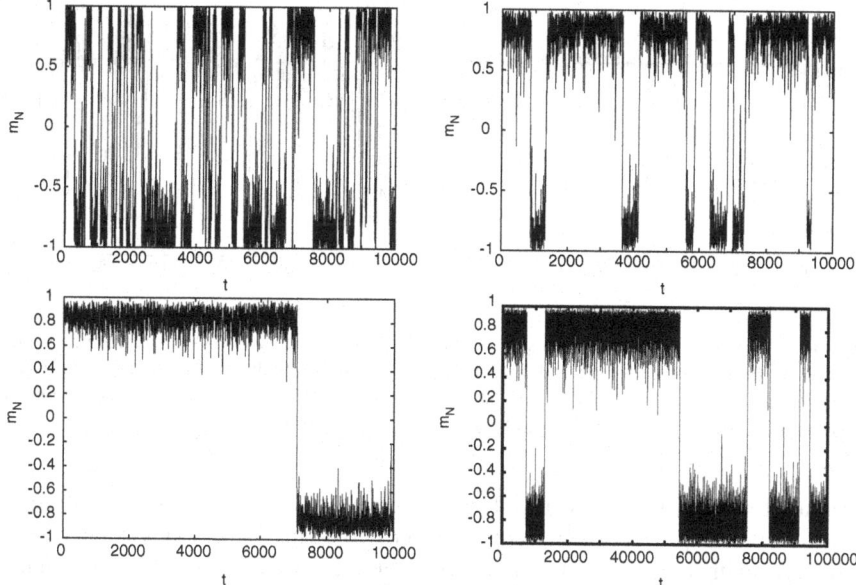

Fig. 10.1 *1st row* magnetisation of a system of $N = 3^3$ spins (*left*) and $N = 4^3$ spins (*right*); *2nd row* magnetisation of a system of $N = 5^3$ spins (*left*) and $N = 5^3$ spins (*right*) but now monitored for 10^5 time-steps. The temperature T in these simulations is chosen as $T = 3.75$, leading to an equilibrium magnetization K in the thermodynamic limit

The simulation may be repeated for larger and larger system sizes; results for the magnetization (10.1) are shown in the left panel of Fig. 10.2 for systems containing $N = 16^3$ and 64^3 spins. The Figure shows that fluctuations of the magnetisation become smaller as the system size is increased. A second effect is that the time spans over which a stable (in the sense of non-switching) magnetisation is observed increases with increasing system size; this is shown for the smaller systems in Fig. 10.1. Indeed, for a system containing $N = 64^3 = 262,144$ spins, transitions between states of opposite magnetization are already so rare[2] that they are out of reach of computational resources available to us, though fluctuations of the magnetization about its average value are still discernible. Note in passing that fluctuations of average properties of a *small subsystem* do *not* decrease if the total system size is increased, and that for the system under consideration they are much larger than those of the entire system as shown in the right panel of Fig. 10.2.

The present example system was set up in a way that states with magnetisations $S_N \simeq \pm 0.83$ would be its only (two) distinct macroscopic manifestations. The simulation shows that transitions between them are observed at irregular time

---

[2] Measured in conventional time units, such a system will exhibit a magnetisation which remains stable for times of the order of several years.

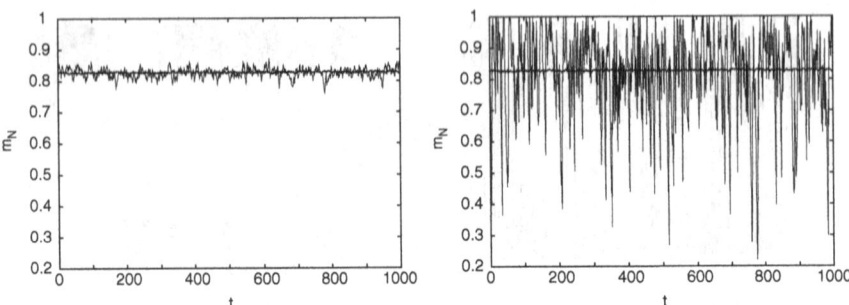

Fig. 10.2 Magnetisation of systems containing $N = 16^3$ and $N = 64^3$ spins showing that fluctuations of the average magnetization of the system decreases with system size (*left panel*), and of a system containing $N = 64^3$ spins, shown together with the magnetization of a smaller subsystem of this system, containing only $N_s = 3^3$ spins (*right panel*)

intervals for small finite systems. This is the macroscopic manifestation of ergodicity. The time span over which a given macroscopic manifestation is stable is seen to increase with system size. It will diverge—ergodicity can be broken—only in the infinitely large system. However, the times over which magnetic properties are stable increase so rapidly with system size that it will be far exceeding the age of the universe for systems consisting of realistically large (though certainly finite) numbers of particles, i.e. for $N = \mathcal{O}(10^{23})$.

Only in the infinite system limit would a system exhibit a strictly constant non-fluctuating magnetisation, and only in this limit would one therefore, strictly speaking, be permitted to talk of a system with a given value of its magnetisation. Moreover, only in this limit would transitions between different macroscopic phases be entirely suppressed and only in this limit could the system therefore, strictly speaking, be regarded as macroscopically stable.

However our simulation already indicates that both properties can be effectively attained in finite (albeit large) systems. The systems just need to be large enough for fluctuations of their macroscopic properties to become sufficiently small as to be practically undetectable at the precision with which these are normally measured, and life-times of different macroscopic phases (if any) must become sufficiently large to exceed all anthropomorphic time scales by sufficiently many orders of magnitude. With respect to this latter aspect of macroscopic stability, reaching times which exceed the age of the universe could certainly be regarded as sufficient for all purposes; these are easily attained in systems of anthropomorphic dimension.

So, there is no explanation of why a finite system would exhibit a strictly constant non-fluctuating magnetisation. Thermodynamics, however, works with of such strictly constant, non-fluctuating properties. This might appear to be a failure of reduction: The properties thermodynamics assumes finite macroscopic systems to have, cannot be explained in terms of the properties of the parts, their interactions. This, however, would be the wrong conclusion to draw. It is essential to note that we do understand two things and these suffice for the behaviour of the compound being reductively explainable: Firstly, we can explain on the basis of the properties of the parts and their interactions why finite systems have a (fairly) stable magnetisation, such that no fluctuations will occur for times exceeding the age of the universe if the systems are sufficiently large. Thus we can explain the observed macro-behaviour reductively. Secondly, we can explain why thermodynamics works even though it uses quantities defined in the thermodynamic limit only: Even though the strictly non-fluctuating properties that thermodynamics works with do not exist in real, i.e. finite systems they are (i) observationally indistinguishable from the properties of finite systems and (ii) we theoretically understand how in the limit $N \to \infty$ fluctuations disappear, i.e. the non-fluctuating properties arise. We would like to argue that this suffices for a reductive explanation of a phenomenon.

In what follows we describe how the suppression of fluctuations of macroscopic quantities in large systems can be understood in terms of statistical limit theorems. We follow a reasoning originally due to Jona-Lasinio (1975) that links these to the coarse grained descriptions and renormalization group ideas, starting in the following section with systems of independent identically distributed random variables, and generalizing in Sect. 10.4 thereafter to systems of interacting degrees of freedom.

10.3 Limit Theorems and Description on Large Scales

Large numbers are according to our reasoning a prerequisite for stability of macroscopic material properties, and only in the limit of large numbers we may expect that macroscopic properties of matter are also non-fluctuating. Early formulations of equations of state of macroscopic systems which postulate deterministic functional relations, e.g. between temperature, density and pressure of a gas, or temperature, magnetisation and magnetic field in magnetic systems can therefore claim strict validity only in the infinite system limit. They are thus seen to presuppose this limit, though in many cases, it seems, implicitly.

From a mathematical perspective, there are—as already stressed by Khinchin in his classical treatise on the mathematical foundations of statistical mechanics (Khinchin 1949)—two limit theorems of probability theory, which are of particular relevance for the phenomena just described: (i) the law of large numbers, according to which a normalized sum of N identically distributed random variables of the form (10.1) will in the limit $N \to \infty$ converge to the expectation $\mu = \langle s_k \rangle$ of the s_k, assuming that the latter is finite; (ii) the central limit theorem according to which the distribution of deviations from the expectation, i.e., the distribution of $S_N - \mu$ for independent identically distributed random variables will in the limit of large N converge to a Gaussian of variance σ^2/N, where $\sigma^2 = \langle s_k^2 \rangle - \langle s_k \rangle^2$ denotes the variance of the s_k.[3]

---

[3] For precise formulations of conditions and proofs, see Feller (1968).

The central limit theorem in particular implies that fluctuations of macroscopic quantities of the form (10.1) will in the limit of large N typically decrease as $1/\sqrt{N}$ with system size N. This result, initially formulated for independent random variables may be extended to so-called weakly dependent variables. Considering the squared deviation $(S_N - \langle S_N \rangle)^2$, one obtains its expectation value

$$\left\langle (S_N - \langle S_N \rangle)^2 \right\rangle = \frac{1}{N^2} \sum_{k,\ell=1}^{N} C_{k,\ell} = \frac{1}{N^2} \sum_{k,\ell=1}^{N} \left\langle (s_k - \langle s_k \rangle)(x_\ell - \langle x_\ell \rangle) \right\rangle \qquad (10.4)$$

and the desired extension would hold for all systems, for which the correlations $C_{k,\ell}$ are decreasing sufficiently rapidly with "distance" $|k - \ell|$ to ensure that the sums $\sum_{\ell=1}^{\infty} |C_{k,\ell}|$ are finite for all k.

The relation between the above-mentioned limit theorems and the description of stochastic systems at large scales are of particular interest for our investigation, a connection that was first pointed out by Jona-Lasinio (1975).[4] The concept of large-scale description has been particularly influential in the context of the renormalization group approach which has led to the our current and generally accepted understanding of critical phenomena.[5]

To discuss this relation, let us return to independent random variables and, generalising Eq. (10.1), consider sums of random variables of the form

$$S_N(s) = \frac{1}{N^{1/\alpha}} \sum_{k=1}^{N} s_k. \qquad (10.5)$$

The parameter α fixes the power of system size N by which the sum must be rescaled in order to achieve interesting, i.e., non-trivial results. Clearly, if the power of N appearing as the normalization constant in Eq. (10.5) is too large for the type of random variables considered (i.e. if α is too small), then the normalized sum (10.5) will almost surely vanish, $S_N \to 0$, in the large N limit. Conversely, if the power of N in Eq. (10.5) too small (α too large), the normalized sum (10.5) will almost surely diverge, $S_N \to \pm\infty$, as N becomes large. We shall in what follows restrict our attention to the two important cases $\alpha = 1$—appropriate for sums of random variables of non-zero mean—and $\alpha = 2$—relevant for sums of random variables of zero mean and finite variance. For these two cases, we shall recover the propositions of the law of large numbers ($\alpha = 1$) and of the central limit theorems ($\alpha = 2$) as properties of *large-scale* descriptions of (10.5).

---

[4] On this, see also Batterman (1998), who referred to the relation on several occasions in the context of debates on reductionism.

[5] The notion of critical phenomena refers to a set of anomalies (non-analyticities) of thermodynamic functions in the vicinity of continuous, second order phase transitions. A lucid exposition is given by Fisher (1983).

To this end we imagine the s_k to be arranged on a linear chain. The sum (10.5) may now be reorganised by (i) combining neighbouring pairs of the original variables and computing their averages $\bar{s}_k$, yielding $N' = N/2$ of such local averages, and by (ii) appropriately rescaling these averages so as to obtain renormalised random variables $s'_k = 2^\mu \bar{s}_k$, and by expressing the original sum (10.5) in terms of a corresponding sum of the renormalised variables, formally

$$\bar{s}_k = \frac{s_{2k-1} + s_{2k}}{2}, \quad s'_k = 2^\mu \bar{s}_k, \quad N' = N/2, \tag{10.6}$$

so that

$$S_N(s) = 2^{(1-1/\alpha-\mu)} S_{N'}(s'). \tag{10.7}$$

One may compare this form of "renormalization"—the combination of local averaging and rescaling—to the effect achieved by combining an initial reduction of the magnification of a microscope (to the effect that only locally averaged features can be resolved) with an ensuing change of the contrast of the image it produces.

By choosing the rescaling parameter μ such that $\mu = 1 - 1/\alpha$, one ensures that the sum (10.5) remains invariant under renormalization,

$$S_N(s) = S_{N'}(s'). \tag{10.8}$$

The renormalization procedure may therefore be iterated, as depicted in Fig. 10.3: $s_k \to s'_k \to s''_k \to \ldots$, and one would obtain the corresponding identity of sums expressed in terms of repeatedly renormalised variables,

$$S_N(s) = S_{N'}(s') = S_{N''}(s'') = \ldots \tag{10.9}$$

The statistical properties of the renormalised variables s'_k will, in general be different from (though, of course, dependent on) those of the original variables s_k, and by the same token will the statistical properties of the doubly renormalised variables s''_k be different from those of the s'_k, and so on. However, one expects that statistical properties of variables will after sufficiently many renormalization steps, i.e., at large scale, eventually become independent of the microscopic details and of the scale considered, thereby becoming largely independent of the statistical properties of the original variables s_k, and invariant under further renormalization. This is indeed what happens under fairly general conditions.

It turns out that for sums of random variables s_k with non-zero average $\mu = \langle s_k \rangle$ the statement of the law of large numbers is recovered. To achieve asymptotic invariance under repeated renormalization, one has to choose $\alpha = 1$ in (10.5), in which case one finds that the repeatedly renormalised variables $s_k^{'''\cdots}$ converge under repeated renormalization to the average of the s_k, which is thereby seen to coincide with the large N limit of the S_N.

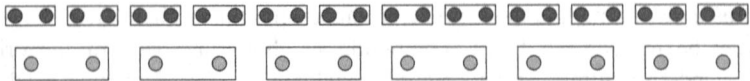

Fig. 10.3 Repeated enlargement of scale: the present example begins with a system of 24 random variables symbolised by the *dots* in the first row. These are combined in pairs, as indicated by *frames* surrounding two neighbouring *dots*. Each such pair generates a renormalised variable, indicated by 12 *dots* of *lighter shade* in the second row of the figure. Two of those are combined in the next step as indicated by *frames* around pairs of renormalised variables, thereby starting the iteration of the renormalization procedure

If sums of random variables of zero mean (but finite variance) are considered, the adequate scaling is given by $\alpha = 2$. In this case the repeatedly renormalised variables $s_k^{''''\cdots}$, and thereby the S_N, are asymptotically normally distributed with variance σ^2 of the original variables s_k, even if these were not themselves normally distributed. The interested reader will find details of the mathematical reasoning underlying these results in Appendix 10.8 below.

Let us not fail to mention that other stable distributions of the repeatedly renormalised variables $s_k^{''''\cdots}$, thus of the S_N—the so-called Lévy α-stable distributions (Feller 1968)—may be obtained by considering sums random variables of infinite variance. Although such distributions have recently attracted some attention in the connection with the description of complex dynamical systems, such as turbulence or financial markets, they are of lesser importance for the description of thermodynamic systems in equilibrium, and we shall therefore not consider these any further in what follows.

Interestingly, the asymptotics of the convergence to the stable distributions under repeated renormalization described above can be analyzed in full analytic detail for the presently considered case of sums of independent random variables. As demonstrated in Appendix 10.9, this allows to quantify the finite size corrections to the limiting distributions in terms of the scaling of high-order cumulants with inverse powers of system size N.

10.4 Interacting Systems and the Renormalization Group

For the purpose of describing macroscopic systems the concept of large-scale descriptions of a system, used above to elucidate the two main limit theorems of mathematical statistics, needs to be generalised to interacting, thus *correlated* or dependent random variables. Such a generalisation was formulated at the beginning of the 1970s as renormalization group approach to interacting systems.

Starting point of this approach is the Boltzmann-Gibbs equilibrium distribution of microscopic degrees of freedom taking the form (10.2). The idea of the renormalization group approach to condensed matter systems is perhaps best explained in terms of the normalisation constant Z_N appearing in (10.2), the so-called partition function. It is related to the dimensionless free energy $\bar{f}_N$ of the system via $Z_N = e^{-N\bar{f}_N}$ and thereby to its thermodynamic functions and properties.[6] To this end the partition function in (10.2) is written in the form

$$Z_N = Z_N(\boldsymbol{K}) = \sum_s e^{-\overline{H}_N(s;\boldsymbol{K})}, \qquad (10.10)$$

in which $\overline{H}_N(s; \boldsymbol{K})$ denotes the dimensionless energy function of the system, i.e., the conventional energy function multiplied by the inverse temperature β, while $\boldsymbol{K}$ stands for the collection of all coupling constants in H_N (multiplied by β). These may include two-particle couplings as in (10.3), but also single-particle couplings as well as a diverse collection of many-particle couplings. Renormalization investigates, how the formal representation of the partition function changes, when it is no longer interpreted as a sum over all micro-states of the original variables, but as a sum over micro-states of renormalised variables, the latter defined as suitably rescaled local averages of the original variables in complete analogy to the case of independent random variables.

In contrast to the case of independent variables, geometric neighbourhood relations play a crucial role for interacting systems, and are determined by the physics of the problem. E.g., for degrees of freedom arranged on a d-dimensional (hyper)-cubic lattice, one could average over the b^d degrees of freedom contained in a (hyper)-cube B_k of side-length b to define locally averaged variables, as illustrated in Fig. 10.4 for $d = 2$ and $b = 2$, which are then rescaled by a suitable factor b^μ in complete analogy to the case of independent random variables discussed above,

$$\bar{s}_k = b^{-d} \sum_{i \in B_k} s_i, \quad s'_k = b^\mu \bar{s}_k. \quad N' = N/b^d. \qquad (10.11)$$

The partition sum on the coarser scale is then evaluated by first summing over all micro-states of the renormalised variables s' and for each of them over all configurations s compatible with the given s', formally

$$Z_N(\boldsymbol{K}) = \sum_{s'} \left[\sum_s P(s', s) e^{-\overline{H}_N(s;\boldsymbol{K})} \right] \equiv \sum_{s'} e^{-\overline{H}_{N'}(s';\boldsymbol{K}')} = Z_{N'}(\boldsymbol{K}') \qquad (10.12)$$

where $P(s', s) \geq 0$ is a projection operator constructed in such a way that $P(s', s) = 0$, if s is incompatible with s', whereas $P(s', s) > 0$, if s is compatible with s', and normalised such that $\sum_{s'} P(s', s) = 1 \forall s$. The result is interpreted as the partition function corresponding to a system of $N' = b^{-d}N$ renormalised variables,

---

[6] The dimensionless free energy is just the product of the standard free energy and the inverse temperature β. At a formal level, the partition function is closely related to the characteristic function of a (set of) random variables, in terms of which we analysed the idea of large-scale descriptions for sums of independent random variables in Appendix 10.8.

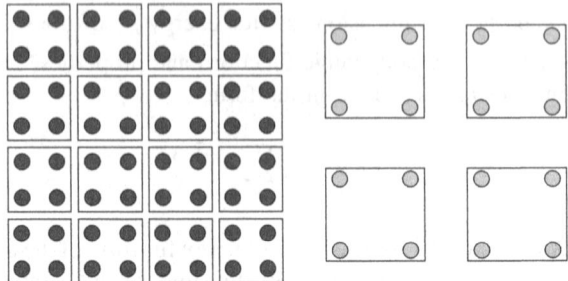

Fig. 10.4 Iterated coarsening of scale

corresponding to a dimensionless energy function $\overline{H}_{N'}$ of the same format as the original one, albeit with renormalised coupling constants $K \rightarrow K'$, as expressed in (10.12). The distance between neighbouring renormalised degrees of freedom is larger by a factor b than that of the original variables. Through an ensuing rescaling of all lengths $\ell \rightarrow \ell/b$ one restores the original distance between the degrees of freedom, and completes the renormalization group transformation as a mapping between *systems of the same format*.

As in the previously discussed case of independent random variables, the renormalization group transformation may be iterated and thus creates not only a sequence of repeatedly renormalised variables, but also a corresponding sequence of repeatedly renormalised couplings

$$K \rightarrow K' \rightarrow K'' \rightarrow K''' \rightarrow \ldots . \tag{10.13}$$

As indicated in Fig. 10.5, this sequence may be visualised as a renormalization group 'flow' in the space of couplings.

The renormalization transformation entails a transformation of (dimensionless) free energies $\overline{f}_N(K) = -N^{-1} \ln Z_N$ of the form

$$\overline{f}_N(K) = b^{-d}\overline{f}_{N'}(K'). \tag{10.14}$$

For the present discussion, however, the corresponding transformation of the so-called correlation length ξ which describes the distance over which the degrees of freedom in the system are statistically correlated, is of even greater interest. As a consequence of the rescaling of all lengths, $\ell \rightarrow \ell/b$, in the final step of the renormalization transformation, one obtains

$$\xi_N(K) = b\xi_{N'}(K'). \tag{10.15}$$

Repeated renormalization amounts to a description of the system on larger and larger length scales. The expectation that such a description would on a sufficiently large scale eventually become independent of the scale actually chosen would

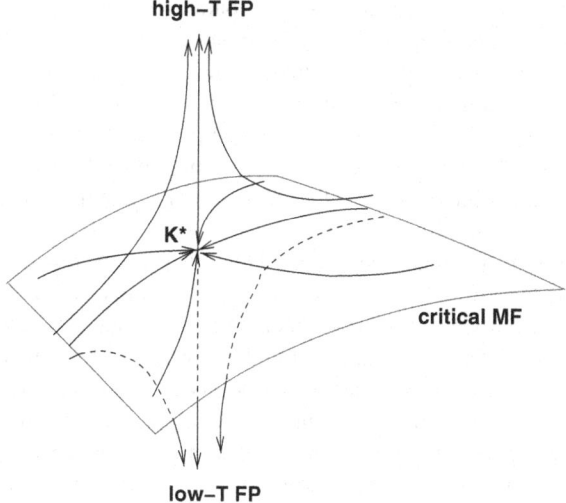

Fig. 10.5 Renormalization group flow in the space of couplings

correspond to the finding that the renormalization group flow would typically approach a fixed point: $K \to K' \to K'' \to \ldots \to K^*$. As in the case of the renormalization of sums of independent random variables exponent μ in the rescaling operation in Eq. (10.11) must be judiciously chosen to allow approach to a fixed point describing non-trivial large scale behaviour.

The existence of fixed points is of particular significance in the limit of infinitely large system size $N = N' = \infty$, as in this limit Eq. (10.15), $\xi_\infty(K^*) = b\xi_\infty(K^*)$ will for $b \neq 1$ *only* allow for the two possibilities

$$\xi_\infty(K^*) = 0 \quad \text{or} \quad \xi_\infty(K^*) = \infty. \tag{10.16}$$

The first either corresponds to a so-called high-temperature fixed point, or to a low-temperature fixed point. The second possibility with infinite correlation length corresponds to a so-called critical point describing a continuous, or second order phase transition. In order to realise the second possibility, the initial couplings K of the system must be adjusted in such a way that they come to lie precisely *on* the critical manifold in the space of parameters, defined as the basin of attraction of the fixed point K^* for the renormalization group flow. If the system's initial couplings K are not exactly *on* the critical manifold, but close to it, repeated renormalization will result in a flow that visits the vicinity of the fixed point K, but is eventually driven away from it (see Fig. 10.5). Within an analysis of the RG transformation that is linearized in the vicinity of K^* this feature can be exploited to analyse critical behaviour of systems in the vicinity of their respective critical points and quantify it in terms of critical exponents, and scaling relations satisfied by them (Fisher 1974, 1983).

The distance from the critical manifold is parameterized by the so-called "relevant couplings".[7] Experience shows that relevant couplings typically form a low-dimensional manifold within the high-dimensional space of system parameters K. For conventional continuous phase transitions it is normally two-dimensional— parametrized by the deviation of temperature and pressure from their critical values in the case of gasses, by corresponding deviations of temperature and magnetic field in magnetic systems, and so forth.[8] The fact that all systems in the vicinity of a given critical manifold are controlled by the same fixed point does in itself have the remarkable consequence that there exist large classes of microscopically very diverse systems, the so-called universality classes, which exhibit essentially the same behaviour in the vicinity their respective critical points (Fisher 1974, 1983).

For the purpose of the present discussion, however, the phenomenology close to *off-critical* high and low-temperature fixed points is of even greater importance. Indeed, all non-critical systems will eventually be driven towards one of these under repeated renormalization, implying that degrees of freedom are virtually *uncorrelated* on large scales, and that the description of non-critical systems within the framework of the two limit theorems for independent variables discussed earlier is therefore *entirely adequate*, despite the correlations over small distances created by interactions between the original microscopic degrees of freedom.

10.5 The Thermodynamic Limit of Infinite System Size

We are ready for a first summary: only in the thermodynamic limit of infinite system size $N \rightarrow \infty$ will macroscopic systems exhibit non-fluctuating thermodynamic properties; only in this limit can we expect that deterministic equations of state exist which describe relations between different thermodynamic properties as well as the manner in which these depend on external parameters such as pressure, temperature or electromagnetic fields. Moreover, only in this limit will systems have strictly stable macroscopic properties in the sense that transitions between different macroscopic phases of matter (if there are any) will not occur in finite time. Indeed stability in this sense is a consequence of the absence of fluctuations, as (large) fluctuations would be required to induce such macroscopic transformations. We have seen that these properties can be understood in terms of coarse-grained

---

[7] These correspond to a subset of the couplings encoded in K, whose distance from the critical manifold is increased under renormalization. The critical manifold itself is parameterized by the so-called irrelevant couplings; their distance from the critical point K^* is decreased under successive renormalizations.

[8] Note that in practical RG analyses of the liquid gas critical point, chemical potential is usually used instead of pressure. Also, proper independent coordinates of the manifold of relevant couplings are not necessarily the physical parameters themselves; they could be, and often are, constructed from suitable combinations thereof. For a detailed discussion, see e.g. Lavis and Bell (1998).

descriptions, and the statistical limit theorems for independent or weakly dependent random variable describing the behaviour averages and the statistics of fluctuations in the large system limit, and we have seen how RNG analyses applied to *off-critical* systems can provide a rationalization for the applicability of these limit theorems.

Real systems are, of course, always finite. They are, however, typically composed of huge numbers of atomic or molecular constituents, numbers so large in fact that they are for the purposes or determining macroscopic physical properties "indistinguishable from infinity": fluctuations of macroscopic properties decrease with increasing system size, to an extent of becoming virtually undetectable in sufficiently large (yet finite) systems. Macroscopic stability (in the sense of absence of transitions between different macroscopic phases on time-scales which exceed the age of the universe) in large finite systems ensues.

One might argue that this is a failure of reduction, and in a sense this is true (as mentioned before). Strictly non-fluctuating properties as postulated or assumed by thermodynamics do only exit in the thermodynamic limit of infinite system size $N \to \infty$. On the basis of finitely many parts of actual systems it cannot be explained why systems have such properties. We suggest that one should bite the bullet and conclude that strictly non-fluctuating properties as postulated by thermodynamics do not exist. But as already mentioned this is less of a problem than it might appear. There are two reasons why this is not problematic. First: We can explain on the basis of the properties of the parts and their interactions of actual finite systems why they have stable properties, in the sense that fluctuations of macroscopic observables will be *arbitrarily* small, and that ergodicity can be broken, with life-times of macroscopically different coexisting phases exceeding the age of the universe. Thus the observed behaviour of the system can be reductively explained. We do have a micro-reduction of the observed behaviour. Second: We furthermore understand how the relevant theories, thermodynamics and statistical mechanics, are related in this case. They are related by the idealization of the thermodynamic limit of infinite system size $N \to \infty$ and we have an account of how de-idealization leads to the behaviour statistical mechanics attributes to finite systems. The stability we observe, i.e. the phenomenon to be explained, is compatible both with the idealization of infinite system size and with the finite, but very large, size of real systems. The fact that strict stability requires the infinite limit poses no problem because we are in a region where the difference between the finite size model and the infinite size model cannot be observed.

The role of the thermodynamic limit and the issue of stability of macroscopic system properties are more subtle, and more interesting, in the case of phase transitions and critical phenomena. To discuss them, it will be helpful to start with a distinction.

What we have discussed so far is the stability of macroscopic physical properties vis-á-vis incessant changes of the system's dynamical state on the micro-level. Let us call this "actual stability" and contrast it with "counterfactual stability". Counterfactual stability is the stability of the macro-behaviour with respect to non-actual counterfactual changes a system's composition at the micro-level might undergo:

e.g., in a ferro-magnetic system one might add next-nearest neighbour interactions to a system originally having only nearest neighbour interactions, and scale down the strength of the original interaction in such a manner that would leave the macroscopic magnetization of a system invariant.

The notion of counterfactual stability applies in particular to the phenomenon of universality of critical phenomena. Critical phenomena comprise a set of anomalies (*algebraic* singularities) of thermodynamic functions that are observed at second order phase transitions in a large variety of systems. Critical phenomena, and specifically the critical exponents introduced to characterize the non-analyticities quantitatively have various remarkable properties. For instance, critical exponents exhibit a remarkable degree of universality. Large classes of systems are characterized by identical sets of critical exponents, despite the fact that interactions at the microscopic level may be vastly different. Within the RNG approach described in the previous section this is understood as a global property of a renormalization group flow: all systems with Hamiltonians described by couplings K in the vicinity of a given critical manifold will be attracted by the same RNG fixed point K^*, and therefore exhibit identical critical exponents.

The case of critical behaviour is thus a special and particularly impressive case of (counterfactual) stability. Note, however, that even though the notion of universality provides a notion of stability (see Batterman 2002, p. 57ff), the range it applies to is fairly restricted and does not cover all the cases Fodor had in mind when he was referring to the stability of macro-level. In particular, universality of critical phenomena as uncovered by the RNG approach *only* refers to asymptotic critical exponents,[9] describing critical singularities only in the *immediate vicinity* of critical points. The thermodynamic properties of a system *not* exactly *at* its critical point will, however, be influenced by the presence of irrelevant couplings, and thus show properties which are system-specific, and *not* universal within universality classes. We will discuss some further notes of caution in Sect. 10.6 below.

The case of critical phenomena requires special discussion also with respect to the infinite system limit. We have seen that the thermodynamic limit is a prerequisite for systems to exhibit non-fluctuating macroscopic physical properties, but that this type of behaviour is well approximated (in the sense of being experimentally indistinguishable from it) in finite sufficiently large systems. Thermodynamically, phase transitions and critical phenomena are associated with non-analyticities in a system's thermodynamic functions. Given certain uncontroversial assumptions, such non-analyticities cannot occur in finite systems (cf. Menon and Callendar 2013, p. 194) in the canonical or grand-canonical ensembles of Statistical Mechanics. For phase transitions to occur, and systems to exhibit critical phenomena it thus appears that an "ineliminable appeal to the thermodynamic limit and to the singularities that emerge in that limit" (Batterman 2011, p. 1038) is required.

---

[9] This includes asymptotic scaling functions and even asymptotic corrections to scaling (Wegner 1976).

Is this an appeal to idealizations that differs from other cases? To discuss this question, we need to return to the finite N versions of the RG flow Eqs. (10.14) and (10.15) for the free energy and the correlation length, respectively. It is customary to use the linear extent L of the system rather than the number of particles $N = L^d$ (assuming hypercubic geometry), to indicate the finite extent of the system, and to rewrite the finite-L RG flow equations in the form

$$\bar{f}_N(K) \equiv \bar{f}(K, L^{-1}) = b^{-d}\bar{f}(K', bL^{-1}) \tag{10.17}$$

and

$$\xi_N(K) \equiv \xi(K, L^{-1}) = b\bar{\xi}(K', bL^{-1}). \tag{10.18}$$

These reformulations already indicate that the inverse L^{-1} of the system's linear dimension L is a *relevant* variable in the RNG sense due to the final rescaling $\ell \to \ell/b$ of all lengths (thus $L^{-1} \to bL^{-1} > L^{-1}$) in the final RG step. The condition for the appearance of this additional relevant variable to be the *only* modification of the RNG transformation is that the system must be sufficiently large that the RG-flow in the space of couplings $K \to K' \to K'' \to \ldots$ is itself unmodified by the finite size of the system. In a real-space picture of RG, it requires in particular all renormalized couplings to be embeddable in the system (for details, see Barber 1983). A finite value of L then implies that a finite system can never be exactly *critical* in the sense of exhibiting an infinite correlation length. As the relevant variable L^{-1} is non-zero, the system is driven away from the critical manifold under renormalization, and indeed coarse graining is impossible beyond the scale L set by the system size. If carried through, this finite-size modification of the RNG transformation gives rise to so-called finite-size-scaling theory (FSS) which describes in quantitative detail the way in which critical singularities are rounded due to finite size effects (Barber 1983). The analysis is more complicated than, but conceptually fully equivalent to the finite size scaling analysis for the statistical limit theorems described in Appendix 10.9. In particular, variables that are irrelevant in the RG sense are scaled with suitable inverse powers of L in finite systems, the powers being related to the eigenvalues of the linearized RG transformation (in complete analogy to the situation described in Appendix 10.9). Thus while proper singularities of thermodynamic functions disappear in systems of finite size, deviations from the behaviour described in terms of the corresponding infinite-system singular functions will become noticeable only in ever smaller parameter regions around "ideal" critical points, as system sizes grow, and will in sufficiently large systems eventually be indistinguishable from it using experiments of conceivably attainable precision. In this sense, the infinite system idealization of singular behaviour of thermodynamic functions in the vicinity of critical points is an idealization, which is controllable in large systems in a well-defined sense, which does not appear to be fundamentally different from that of non-fluctuating thermodynamic functions and absolute stability discussed earlier.

As before there is no explanation of why a finite system would exhibit phase transitions in the strict sense. Phase transitions as defined by thermodynamics do only exist in the thermodynamic limit of infinite system size $N \rightarrow \infty$. On the basis of finitely many parts of actual systems it cannot be explained why systems have such properties. The same applies to the universal behaviour of systems in the same universality classes. Strictly speaking universality only obtains in the thermodynamic limit. Neither the occurrence of phase transitions nor universality can be explained in terms of the properties of the parts, their interactions. This might appear to be a failure of reduction. That would, however, be the wrong conclusion to draw. Again, in both cases (the occurrence of phase transitions and the universal behaviour at the critical point) we do understand two things and these suffice for the behaviour being reductively explainable: Firstly, in the case of phase-transitions, we can explain on the basis of the properties of the parts and their interactions why finite systems exhibit behaviour that is observationally indistinguishable from strict phase-transitions, which involve non-analyticities (For this point see also Kadanoff 2013, p. 156 or Menon and Callendar 2013, pp. 212–214). Thus we can explain the observed macro-behaviour reductively. In the case of universality, finite-size-scaling theory makes available reductive explanations of the observed similarities in the macro-behaviour of different kinds of systems. Secondly, we can explain why thermodynamics works as well as it does even though it uses quantities defined in the thermodynamic limit only: Even though neither phase transitions as defined in thermodynamics do not exist in real, i.e. finite systems nor the phenomenon of universality (in the strict sense), they are (i) observationally indistinguishable from the properties and behaviour of finite systems and (ii) we theoretically understand how in the limit $N \rightarrow \infty$ phase transitions and universal behaviour (in the strict sense) would arise. In short: We idealize, but we understand how the idealizations work (For a discussion of some of these points see also Butterfield 2011, Sect. 7 as well as Menon and Callender 2013, Sect. 3.2).

10.6 Supervenience, Universality and Part-Whole-Explanation

In the previous section we have argued that we do have reductive explanations of the observed behaviour that in thermodynamics is described in terms of phase transitions and universal behaviour. In this section we will deal with a possible objection. It might be argued that for a reductive explanation of the macro-behaviour the properties of the constituents have to determine the properties of the compound. However, in the cases we are discussing, no such determination relation obtains. In this section we would like to reject this claim.

When the macro behaviour of physical systems is stable, many details of the exact micro-state are irrelevant. This is particularly impressive in the case of universality.

Margret Morrison argues that if the microphysical details are irrelevant, the phenomenon in question is not reducible to its microphysical constituents and that the supervenience relation is "inapplicable in explaining the part-whole aspects" of such phenomena (Morrison 2012, p. 156). Morrison classifies universal behaviour at phase transitions as "emergent", characterizing emergence as follows:

> ...what is truly significant about emergent phenomena is that we cannot appeal to micro-structures in explaining or predicting these phenomena, even though they are constituted by them (Morrison 2012, p. 143).

Even though we agree with Morrison that universal phenomena are in an interesting sense ontologically independent of the underlying microstructure we reject her claim that this violates the "reductionist picture" (Morrison 2012, p. 142). We will focus on one line of her argument. What Morrison calls the "reductionist picture" entails the claim that the constituents properties and states determine the behaviour of the compound system. The reductionist picture entails (or presupposes) the supervenience of the properties of the compound on the properties (and interactions) of the parts. Only if the constituents' properties and interactions determine the compounds properties can we reductively explain the latter in terms of the former. One problem for the reductionist, according to Morrison, is the failure of supervenience.

Why would one suppose that supervenience fails in the case of stable macro-behaviour and universal behaviour in particular? In Morrison's paper we find two arguments. The first has to do with the thermodynamic limit. The stable macro-behaviour to be explained presupposes an infinite number of constituents. Real systems are finite. The behaviour to be explained is not determined by the behaviour of the finite number of constituents. We have already dealt with this issue in the previous section and have indicated why we are not convinced by this line of argument. We thus move to her second argument:

> If we suppose that micro properties could determine macro properties in cases of emer-gence, then we have no explanation of how universal phenomena are even possible. Because the latter originate from vastly different micro properties, there is no obvious ontological or explanatory link between the micro and macro levels (Morrison 2012, p. 162).

We fail to see why it should be impossible that vastly different micro-properties determine the same micro-property. To illustrate: the integer 24 may be obtained as a sum of smaller integers in many different ways $(24 = 13 + 11 = 10 + 14 = 9 + 8 + 7$ etc.). However, this is not a valid argument for the claim that the summands fail to determine the sum. Similarly, the fact that a multiplicity of micro-states gives rise to the same macro-state is no objection to the claim that the micro-state determines the macro-state.

In fact, the simulation in Sect. 10.2 and the explanations in Sects. 10.3 and 10.4 aim at explaining how this is possible. The simulation in Sect. 10.2 has *illustrated* that, by simply increasing the sample size, fluctuations decrease and macroscopic modifications become more and more rare, i.e. the macro-state becomes more and more stable despite of changes in the micro-states. In Sect. 10.3 we discussed an *explanation* for this phenomenon for the case of non-interacting constituents/variables. It is in virtue of the central limit theorem that fluctuations of macroscopic observables decrease with the number N of constituents as $\frac{1}{\sqrt{N}}$ for large system sizes, and that stable behaviour exists for large systems. In Sect. 10.4 we moved to the case of interacting variables/constituents. Again, what is provided is an *explanation* for why so many features of the micro-system are irrelevant. RNG explains how systems that are characterized by very different Hamilton operators nevertheless give rise to very similar macro-behaviour. This only works because, given a particular Hamiltonian, i.e. the micro-properties, the macro-behaviour is fixed and thus determined. If supervenience would indeed fail it would be indeterminate how the Hamiltonian (which represents the properties of the constituents and their interaction) would behave in phase space under renormalization. The RNG-theory explains universality by showing that a whole class of Hamiltonians is attracted by the same fixed point under iterated renormalizations. If the macro-behaviour of the systems in question would not supervene on the micro-structure, the RNG-explanation would not get started.

These explanations tell us *why* and *in which sense* certain features of the constituents or their interactions become irrelevant: note that irrelevance in the RNG sense acquires a technical meaning, which coincides with the plain-English meaning of the word only as far as the determination of asymptotic critical exponents (in infinitely large systems) is concerned.

The fact that certain features of the constituents are irrelevant in the technical RNG sense does therefore not imply that the properties and states of the constituents fail to influence the macro-behaviour. Rather, it is only a small number of features of these that does the work for asymptotic critical exponents. Interestingly, these same features are also responsible for driving an off-critical system away from critical RNG fixed points and towards one of the fixed-points at which the (coarse-grained) system appears as a collection of *uncorrelated* degrees of freedom, for which the limit theorems for uncorrelated or weakly correlated random variables provide an appropriate description.

Whenever a system is not exactly at its critical point, there will always be a residual effect of the so-called irrelevant variables on thermodynamic behaviour. A *finite* system, in particular, is never exactly at a critical point, as $1/L$ (with L denoting the linear dimension of the system) is always a relevant variable in the renormalization group sense (increasing under renormalization/coarse graining); this leads to rounding of critical singularities, which can be quantitatively analysed (see Sect. 10.5). As a consequence, Morrison's argument, if valid, would not apply to finite systems (nor to systems which are off-critical for other reasons), and therefore the so-called irrelevant variables will contribute to the determination of the macro-behaviour of all finite systems.

It may be worth at this point to explicitly discuss the specific nature and reach of RNG analyses of macroscopic systems. To begin with it is helpful to recall that with respect to the analysis of collective behaviour in general and of phase transitions in particular RNG has a *twofold* explanatory role. These two roles concern two sets of questions that ought to be distinguished.

The first set of questions addressed by RNG concerns the behaviour of individual critical systems: Why does the compressibility of water diverge according to some fixed function of temperature etc.? With regard to questions like this RNG is simply a coarse-graining procedure that allows us to calculate approximately correct results. The explanation of the single system's critical exponents starts with the description of the microstructure. In this context RNG is effectively used as a tool for micro-reductive explanation of the system's behaviour. RNG is merely an effective tool to evaluate thermodynamic functions of interacting systems (albeit in the majority of cases only approximately), where exact evaluations are infeasible. In some sense, RNG can be regarded as a successor theory of mean field theory, which was unable to produce even approximately correct critical exponents.

There is, however, a second set of questions. Micro-reductive explanations may appear to be unable to answer these. These questions concern universality and scaling (i.e. the fact that the critical exponents obey certain system-independent relations). Why are there universality classes at all, i.e. why is it that systems with extremely different microstructure such as alloys ferro-magnets and liquids obey exactly the same laws near their critical points, i.e. why is it that the values of the critical exponents of the members of such classes coincide? Furthermore, why is it the case that all critical systems—irrespective their universality class—obey the scaling relations?[10]

RNG appears to answer the above questions non-reductively. The essential ingredients of such an explanation are not properties of the constituents of single critical systems, one might argue, but rather properties of the renormalization-group-flow—topological features of the space of coupling constants (see Morrison 2012, p. 161/2). The renormalization-group-transformation induces a mapping between Hamiltonians or the coupling-constants of the Hamiltonians in question. The iteration of such a mapping defines a flow in the space of coupling constants as represented in Fig. 10.5. Fixed points come along with basins of attraction. All Hamiltonians/physical systems within such a basin flow towards the same critical point, i.e. their behaviour is identical. It is —basically—the existence of these basins of attraction that explain why physical systems with widely diverging microstructure behave identically near the critical point.

The explanation of universality and the scaling relation do not appear to appeal to the microscopic constitution of individual systems that show this behaviour. Instead it appeals to topological features of the space of coupling constants/ Hamiltonians.

---

[10] For the sake of completeness, we note that there are well-understood exceptions (see e.g. Wegner 1976), which we need, however, not discuss in the context of the present paper.

But the reductionist is able to defend her view: Why should micro-explanations be unable to explain that a variety of systems behave identical near the critical point? There is a feature on the micro-level that all of these systems have in common. And it is in virtue of this common feature that RNG works:

> The starting point in the renormalization group approach is to realize that the most important fluctuations at the critical point have no characteristic length. Instead the important fluctuations have *all* wavelengths ranging from the atomic spacing up to the correlation length; close to the critical point the correlation length is much larger than the atomic spacing (Wilson 1976).

The absence of a characteristic length (the divergence of the correlation length) *at* the critical point implies invariance of a system under coarse graining, or scale transformations. An RNG transformation which describes the effect of coarse graining in terms of a transformation $K \rightarrow K'$ of the systems' set of couplings will therefore identify a critical point with a fixed point K^* of that transformation. The reductionist will thus think of topological features of the transformation such as basins of attraction (i.e. universality classes) as an unavoidable consequence of this type of analysis.

This is finally, once more the point to recall that there is a second way in which fluctuations of thermodynamic properties of a system can fail to exhibit a characteristic length: this is the situation where the correlation length *vanishes* and the system is statistically fully homogeneous (and therefore also scale invariant). This possibility is realized at the high- or low-temperature fixed points of the RNG transformations (as discussed in Sect. 10.4), which will come with *their own* basins of attraction and associated notions of insensitivity to detail, as embodied in statistical limit theorems.

To sum up: What we claim to have shown is that the reductionist picture according to which the constituents' properties and states *determine* the behaviour of the compound system, and the macro-phenomena can be explained in terms of the properties and states of the constituents is neither undermined by stable phenomena in general nor by universal phenomena in particular.

Let us add, however, that the reductionist picture as outlined above does not imply that an explanation of the macro-behaviour of a system always has to appeal to the micro-level. The fact that it is possible to give such an explanation does not imply that it is the best explanation let alone the only available explanation. In fact, once we know that many details are irrelevant, we have a perfectly good reason to focus on those factors that are not irrelevant. This is the topic of the following section.

10.7 Post Facto Justification of Modelling

Let us close with the following observation: A remarkable feature of our analysis of stability of macroscopic system properties based on a renormalization group approach is the fact that it provides us with a justification for certain kinds of idealisation. Models that are used to describe critical (or other collective) behaviour of given physical systems are often grossly simplified, and it is a non-trivial problem to understand why such models can nevertheless be as successful, as they apparently are. RNG illuminates this point as follows. It teaches us that systems, which are described by Hamiltonians that differ only with respect to irrelevant couplings (systems within the basin of attraction of the RNG fixed point relevant for the large scale phenomenon under study) will under renormalization be attracted to the same fixed point, and will therefore exhibit the same type of collective behaviour behaviour. The presence of irrelevant couplings can in this sense be regarded as belonging to the inessential details which do not significantly affect the problem under study, such as critical behaviour of a given many-particle system, or macroscopic off-critical properties. In a quantitatively well-defined sense, such irrelevant couplings can therefore be neglected when analysing collective phenomena in such systems—critical or not. A description of a system which neglects or abstracts from these details constitutes what would properly be described as an idealized description. Within the RNG setting we therefore have a well-defined and even quantifiable notion of the sense in which an idealized description of a complex system will nevertheless capture the essence of its collective behaviour.

It is a remarkable fact that conventional second order phase transitions are characterized by just two relevant couplings in the RNG sense; this tells us that all but two operators within an infinite set characterizing the universe of possible Hamiltonians for a given system are irrelevant, and can therefore be neglected when analysing asymptotic critical behaviour associated with second order phase transitions. Moreover it is the presence of these same two relevant couplings which is responsible for driving the RNG flow to off-critical fixed points at which systems are statistically fully homogeneous and their macroscopic properties thus describable in terms of statistical limit theorems for uncorrelated or weakly correlated random variables.

It is due to this remarkable fact that simplified models do at all allow us to capture the essence of collective behaviour. In this sense RNG provides a justification for the idea of modelling per se, and it gives us a glimpse on the reasons why "simple" models of condensed matter could at all be successful. Note finally that it is essential that systems are large to allow levels of description which are sufficiently coarse-grained compared to atomistic scales. Stability, and as we have just discussed, simplicity only arises at this sufficiently coarse scale. It has been pointed out (Kühn 2008) that in this sense large numbers appear to be an essential prerequisite for facilitating successful theory formation for condensed matter systems.

A.1 Renormalization and Cumulant Generating Functions

The renormalization group transformation for the case of sums of independent random variables is best investigated in terms of their *cumulant generating functions*.

Given a random variable X, its characteristic function is defined as the Fourier transform of its probability density p_X,[11]

$$\varphi_X(k) = \langle e^{ikX} \rangle = \int dx\, p_X(x)\, e^{ikx}. \tag{10.19}$$

Characteristic functions are important tools in probability. Among other things, they can be used to express moments of a random variable in compact form via differentiation,

$$\left(-i\frac{d}{dk}\right)^n \varphi_X(k)|_{k=0} = \langle X^n \rangle = \int dx\, p_X(x)\, x^n. \tag{10.20}$$

For this reason, characteristic functions are also referred to as *moment generating functions*. A second important property needed here is that the characteristic function of a sum $X + Y$ of two independent random variables X and Y is given by the product of their characteristic functions. For, denoting by p_X and p_Y the probability densities corresponding to the two variables, one finds

$$\varphi_{X+Y}(k) = \int dxdy\, p_X(x)p_Y(y)\, e^{ik(x+y)} = \varphi_X(k)\varphi_Y(k). \tag{10.21}$$

Rather than characterizing random variables in terms of their moments, it is common to use an equivalent description in terms of so-called cumulants instead. Cumulants are related to moments of centered distributions and can be seen to provide measures of "dispersion" of a random variable. They are defined as expansion coefficients of a cumulant-generating function (CGF) which is itself defined as the logarithm of the moment generating function

$$f_X(k) = \log \varphi_X(k); \tag{10.22}$$

hence n-th order cumulants $\kappa_n(X)$ are given by

$$\kappa_n(X) \equiv \left(-i\frac{d}{dk}\right)^n f_X(k)|_{k=0}, \quad n \geq 1. \tag{10.23}$$

---

[11] In this appendix, we follow the mathematical convention to distinguish in notation between a random variable X and its realisation x.

The two lowest order cumulants are $\kappa_1(X) = \mu$, and $\kappa_2(X) = \mathrm{Var}(X)$, i.e., the mean and the second centered moment. The multiplication property of characteristic functions of sums of independent random variables translates into a corresponding addition property of the CGF of sums of independent variables,

$$f_{X+Y}(k) = f_X(k) + f_Y(k), \qquad (10.24)$$

entailing that cumulants of sums of independent random variables are additive.

We note in passing that characteristic functions are the probabilistic analogues of partition functions in Statistical Mechanics, and hence that cumulant generating functions are probabilistic analogues of free energies.

We now proceed to use the additivity relations of CGFs to investigate the properties of sums of random variables (10.5),

$$S_N(s) = \frac{1}{N^{1/\alpha}} \sum_{k=1}^{N} s_k$$

under renormalization. We denote CGF corresponding to S_N by F_N. Let f_1 denote the CGF of the original variables, f_2 that of the renormalised variables s'_k (constructed from sums of two of the original variables), and more generally, let f_{2^ℓ} denote the CGF of the ℓ-fold renormalised variables, constructed from sums involving 2^ℓ original variables. We then get

$$
\begin{aligned}
F_N(k) :&= \log\langle e^{ikS_N}\rangle = Nf_1\left(\frac{k}{N^{1/\alpha}}\right) = \frac{N}{2}f_2\left(\frac{k}{(N/2)^{1/\alpha}}\right) \\
&= \frac{N}{4}f_4\left(\frac{k}{(N/4)^{1/\alpha}}\right) = \cdots = \frac{N}{2^\ell}f_{2^\ell}\left(\frac{k}{(N/2^\ell)^{1/\alpha}}\right).
\end{aligned}
\qquad (10.25)
$$

Assuming that multiply renormalised variables will acquire asymptotically stable statistical properties, i.e. statistical properties that remain invariant under further renormalization, the f_{2^ℓ} would have to converge to a limiting function f^*,

$$f_{2^\ell} \to f^*, \quad \ell \to \infty. \qquad (10.26)$$

This limiting function f^* would have to satisfy a functional *self-consistency* relation of the form

$$f^*(2^{1/\alpha}k) = 2f^*(k) \qquad (10.27)$$

which follows from (10.25). This condition states that the invariant CGF $f^*(k)$ must be a *homogeneous function* of degree α.

The solutions of this self-consistency relation for $\alpha = 1$ and $\alpha = 2$ are thus seen to be given by

$$f^*(k) \equiv \lim_{N \to \infty} F_N(k) = c_\alpha k^\alpha, \tag{10.28}$$

One identifies the CGF of a non-fluctuating (i.e. constant) random variable with $c_\alpha = i\langle X \rangle = i\mu$ for $\alpha = 1$, and that of a Gaussian normal random variable with zero-mean and variance σ^2 with $c_\alpha = -\frac{1}{2}\sigma^2$ for $\alpha = 2$, and thereby verifies the statements of the two limit theorems.

One can also show that the convergence (10.26) is realised for a very broad spectrum of distributions for the microscopic variables, both for $\alpha = 1$ (the law of large numbers), and for $\alpha = 2$ (the central limit theorem). For $\alpha = 1$, there is a "marginal direction" in the infinite-dimensional space of possible perturbations of the invariant CGF (corresponding to a change of the expectation value of the random quantities being summed), which doesn't change its distance to the invariant function $f^*(k)$ under renormalization. All other perturbations are irrelevant in the sense that their distance from the invariant CGF will diminish under repeated renormalization. For $\alpha = 2$ there is one "relevant direction" in the space of possible perturbations, in which perturbations of the invariant CGF will be amplified under repeated renormalization (it corresponds to introducing a non-zero mean of the random variables being added), and a marginal direction that corresponds to changing the variance of the original variables. All other perturbations are irrelevant and will be diminished under renormalization. The interested reader will find a formal verification of these statements in the following Appendix 10.9. Interestingly that stability analysis will *also* allow to quantify the rate of convergence to the limiting distribution as a function of system size N (for sums of independent random variables which—apart from having finite cumulants—are otherwise arbitrary).

A.2 Linear Stability Analysis

Statements about the stability of invariant CGF under various perturbations are proved by looking at the linearisation of the renormalization group transformation in the vicinity of the invariant CGF. We shall see that this description considerably simplifies the full analysis compared to the one in terms of probability densities used in (Sinai 1992).

Let R_α denote the renormalization transformation of a CGF for the scaling exponent α. From (10.25), we see that its action on a CGF f is defined as

$$R_\alpha[f](2^{1/\alpha}k) = 2f(k). \tag{10.29}$$

Assuming $f = f^* + h$, where h is a small perturbation of the invariant CGF, we have

$$R_\alpha[f^* + h](2^{1/\alpha}k) = 2(f^*(k) + h(k)). \tag{10.30}$$

Using an expansion of the transformation R_α in the vicinity of f^*, and denoting by $D_\alpha = D_\alpha[f^*]$ the operator of the linearised transformation in the vicinity of f^* on the l.h.s, one has $R_\alpha[f^* + h] \simeq R_\alpha[f^*] + D_\alpha h$ to linear order in h, thus

$$R_\alpha[f^*](2^{1/\alpha}k) + D_\alpha h(2^{1/\alpha}k) \simeq 2f^*(k) + 2h(k). \tag{10.31}$$

By the invariance of f^* under R_α, we get

$$D_\alpha h(2^{1/\alpha}k) = 2h(k) \tag{10.32}$$

to linear order. The stability of the invariant CGF is then determined by the *spectrum* of D_α, found by solving the *eigenvalue problem*

$$D_\alpha h(2^{1/\alpha}k) = 2h(k) = \lambda h(2^{1/\alpha}k). \tag{10.33}$$

Clearly this equation is solved by homogeneous functions:

$$h(k) = h_n(k) = \kappa_n \frac{(ik)^n}{n!}, \tag{10.34}$$

for which

$$2h_n(k) = \lambda_n h_n(2^{1/\alpha}k)$$

entails

$$\lambda_n = 2^{1-n/\alpha}. \tag{10.35}$$

In order for $f^* + h_n$ to describe a system with finite cumulants, we must have $n \geq 1$.

For the case $\alpha = 1$ then we have $\lambda_1 = 1$ (the corresponding perturbation being marginal), and $\lambda_n < 1$ for $n > 1$ (the corresponding perturbations thus being irrelevant). The marginal perturbation amounts to changing the mean of the random variable to $\mu + \kappa_1$, as mentioned earlier.

In the case where $\alpha = 2$ we have that $\lambda_1 = 2^{\frac{1}{2}}$ (the corresponding perturbation being relevant), $\lambda_2 = 1$ (the corresponding perturbation being marginal), and $\lambda_n < 1$

for all $n > 2$ (the corresponding perturbations thus being irrelevant). The relevant perturbation amounts to introducing a nonzero mean $\mu = \kappa_1$ of the original random variables, while the marginal perturbation changes the variance to $\sigma^2 + \kappa_2$, as mentioned earlier. All other perturbations change higher order cumulants of the random variables considered and are irrelevant.

Knowledge about the eigenfunctions of the linearized RG transformation and their eigenvalues allows to obtain a complete overview over the finite N corrections to the limit theorems we have looked at in Appendix 10.8. Suppose we have

$$f_1(k) = f^*(k) + \sum_{n>1} h_n(k) = f^*(k) + \sum_{n>1} \kappa_n \frac{(ik)^n}{n!}$$

in (10.25). Then after ℓ coarse-graining steps we have

$$\begin{aligned} F_N(k) &= N f_1\left(\frac{k}{N^{1/\alpha}}\right) = f^*(K) + N \sum_{n>1} h_n\left(\frac{k}{N^{1/\alpha}}\right) \\ &= f^*(K) + \frac{N}{2^\ell} \sum_{n>1} \lambda_n^\ell h_n\left(\frac{k}{(N/2^\ell)^{1/\alpha}}\right) \end{aligned} \tag{10.36}$$

where we have exploited the invariance and homogeneity of $f^*(k)$, and the fact that each coarse graining step rescales the eigenfunction h_n by an eigenvalue λ_n. We have recorded this relation in a slightly more complicated version than necessary to formally link it up with the analogous steps used in the derivation of finite-size scaling relations in the case of interacting systems. The reader is invited to check correctness of (10.36) herself using nothing but the homogeneity of the h_n.

In a system with finite N, the number ℓ of coarse graining steps that can be performed is necessarily finite, and in fact restricted to $2^{\ell_{max}} = N$. Using this maximum value in (10.36), we get

$$F_N(k) = f^*(K) + \sum_{n>1} \lambda_n^{\ell_{max}} h_n(k) = f^*(K) + \sum_{n>1} N^{1-n/\alpha} h_n(k) \tag{10.37}$$

which is the result that would have been obtained by just using homogeneity in $F_N(k) = N f_1\left(\frac{k}{N^{1/\alpha}}\right)$. This result entails that higher order cumulants of S_N scale with inverse powers of N according to

$$\kappa_n(S_N) = N^{1-n/\alpha} \kappa_n(S_1), \tag{10.38}$$

so that, e.g., all cumulants higher than the second order cumulant will vanish in the infinite system limit in the case $\alpha = 2$ of the central limit theorem.

References

Barber, M.N.: Finite-size scaling. In: Domb, C., Lebowitz, J.L. (eds.) Phase Transitions and Critical Phenomena, vol. 8, pp. 146–266. Academic Press, London (1983)

Batterman, R.: Why equilibrium statistical mechanics works: universality and the renormalization group. Philos. Sci. **65**, 183–208 (1998)

Batterman, R.: Multiple realizability and universality. Br. J. Philos. Sci. **51**, 115–145 (2000)

Batterman, R.: The Devil in the Details. Oxford University Press, Oxford (2002)

Batterman, R.W.: Emergence, singularities, and symmetry breaking. Found. Phys. **41**, 1031–1050 (2011)

Bedau, M., Humphreys, P.: Introduction. In: Bedau, M., Humphreys, P. (eds.) Emergence: Contemporary Readings in Philosophy and Science, pp. 1–6. Bradford Books, Cambridge (2008)

Binder, K., Stauffer, D.: A simple introduction to monte carlo simulation and some specialised topic. In: Binder, K. (ed.) Applications of the Monte Carlo Method in Statistical Physics, pp. 1–36, 2nd edn. Springer, Berlin (1987)

Butterfield, J.: Less is different: emergence and reduction reconciled. Found. Phys. **41**, 1065–1135 (2011)

Feller, W.: An Introduction to Probability Theory and its Applications, vols. I and II, 3rd edn. Wiley, New York (1968)

Fisher, M.E.: The renormalization group in the theory of critical behavior. Rev. Mod. Phys. **46**, 597–616 (1974)

Fisher, M.E.: Scaling, universality and renormalization group theory. In: Hahne, F.J.W. (ed.) Critical Phenomena: Proceedings of the Summer School Held at the University of Stellenbosch, South Africa, January 1829, 1982. Springer Lecture Notes in Physics, vol. 186, pp. 1–139. Springer, Berlin (1983)

Fodor, J.: Special sciences: still autonomous after all these years. Philos. Perspect. **11**, 149–163 (1997)

Jona-Lasinio, G.: The renormalization group: a probabilistic view. In: Il Nuovo Cimento **26B**(1), 99–119 (1975)

Kadanoff, L.: Theories of matter: infinities and renormalization. In: Robert Batterman (ed.) The Oxford Handbook of Philosophy of Physics, Oxford, 141–188 (2013)

Khinchin, A.I.: Mathematical Foundations of Statistical Mechanics. Dover, New York (1949)

Kühn, R.: Über die konstitutive Rolle des Undendlichen bei der Entstehung physicalischer Theorien füer makroskopische Systeme. In: Brachtendorf, J., Möllenbeck, T., Nickel, G., Schaede, S. (eds.) Unendlichkeit. Mohr Siebeck, Tübingen (2008). http://www.mth.kcl.ac.uk/~kuehn/published/Infinity.pdf

Lange, M.: Laws and Lawmakers. Oxford University Press, Oxford (2009)

Laughlin, R., Pines, D.: The theory of everything. In: Proceedings of the National Academy of Sciences, vol. 97 (2000) (reprinted In: Bedau, M., Humphreys, P. (eds.) Emergence: Contemporary Readings in Philosophy and Science. Bradford Books, Cambridge, pp. 259–268 (2008))

Lavis, D.A., Bell, G.M.: Statistical Mechanics of Lattice Systems, vol. 2. Springer, Berlin (1998)

Menon, T., Callender, C.: Turn and face the strange Ch-Ch-changes: philosophical questions raised by phase transitions. In: Batterman, R. (ed.) The Oxford Handbook of Philosophy of Physics. Oxford University Press, Oxford, pp. 189–223 (2013)

Mitchell, S.: Biological Complexity and Integrative Pluralism. Cambridge University Press, Cambridge (2003)

Morrison, M.: Emergent physics and micro-ontology. Philos. Sci. **79**(1), 141–166 (2012)

Rudder-Baker, L.: Non-reductive materialism. In: McLaughlin, B.P., Beckermann, A., et al. (eds.) The Oxford Handbook of Philosophy of Mind. Oxford University Press, Oxford, pp. 109–127 (2009)

Sinai, Y.: Probability Theory: an Introductory Course, Berlin (1992)

Wegner, F.: The critical state, general aspects. In: Domb, C., Green, M.S. (eds.) Phase Transitions and Critical Phenomena, vol. 6, pp. 7–124. Academic Press, London (1976)

Wilson, K.G.: The renormalization group—introduction. In: Domb, C., Green, M.S. (eds.) Phase Transitions and Critical Phenomena, vol. 6, pp. 1–5. Academic Press, London (1976)

Woodward, J.: Causation with a human face. In: Corry, R., Huw, P. (eds.) Causation, Physics, and the Constitution of Reality. Russells Republic Revisited, pp. 66–105. Oxford University Press, Oxford (2007)

Chapter 11
Between Rigor and Reality: Many-Body Models in Condensed Matter Physics

Axel Gelfert

11.1 Introduction

Scientific models are increasingly being recognized as central to the success and coherence of scientific practice. In the present paper, I focus on a particular class of models intended to describe and explain the physical behaviour of systems that consist of a large number of interacting particles. Such *many-body models*, usually characterized by a specific Hamiltonian (energy operator), are frequently employed in condensed matter physics in order to account for phenomena such as magnetism, superconductivity, and other phase transitions. Because of the dual role of many-body models as models *of physical systems* (with specific physical phenomena as their explananda) as well as mathematical structures, they form an important sub-class of scientific models, from which one can expect to draw general conclusions about the function and functioning of models in science, as well as to gain specific insight into the challenge of modelling complex systems of correlated particles in condensed matter physics. Throughout the present paper, equal emphasis is placed on the process of constructing models and on the various considerations that enter into their evaluation.

The rest of this paper is organized as follows. In the next section, I place many-body models in the context of the general philosophical debate about scientific models (especially the influential 'model as mediators' view), paying special attention to their status as mathematical models. Following this general characterization, the Sect. 11.3 discusses a number of historical examples of many-body models and the uses to which they have been put in 20th-century physics, not least in the transition from classical models of interacting particles to a full appreciation of the quantum aspects of condensed matter phenomena. On the basis of these

A. Gelfert (✉)
Department of Philosophy, National University of Singapore,
10 Kent Ridge Crescent, Singapore 119260, Republic of Singapore
e-mail: axel@gelfert.net

© Springer-Verlag Berlin Heidelberg 2015
B. Falkenburg and M. Morrison (eds.), *Why More Is Different*,
The Frontiers Collection, DOI 10.1007/978-3-662-43911-1_11

historical examples, the next Sect. 11.4 distinguishes between different strategies of model construction in condensed matter physics. Contrasting many-body models with phenomenological models (which are typically derived by interpolating between specific empirical phenomena), it is argued that the construction of many-body models may proceed either from theoretical 'first principles' (sometimes called the ab initio approach) or may be the result of a more constructive application of the formalism of many-body operators. This formalism-based approach, it is argued in the section that follows Sect. 11.5, leads to novel theoretical contributions by the models themselves (one example of which are so-called 'rigorous results'), which in turn gives rise to cross-model support between models of different origins and opens up prospects for exploratory uses of models with a view to fostering model-based understanding. The paper concludes with an appraisal of many-body models as a specific way of investigating condensed matter phenomena that steers a middle path 'between rigor and reality'.

11.2 Many-Body Models as Mathematical Models

Among the various kinds of models used in condensed matter physics, an important subclass are *many-body models* which represent a system's overall behaviour as the collective result of the interactions between its constituents. The present section discusses many-body models in general terms, situating them within the general philosophical debate about scientific models and, more specifically, discussing their status as mathematical models.

Mathematical models can take different forms and fulfill different purposes. They may be limiting cases of a more fundamental, analytically intractable theory, for example when modelling planetary orbits as if planets were independent mass points revolving around an infinitely massive sun. Sometimes models connect different theoretical domains, as is the case in hydrodynamics, where Prandtl's boundary layer model interpolates between the frictionless 'classical' domain and the Navier–Stokes domain of viscous flows (Morrison 1999). Even where a fundamental theory is lacking, mathematical models may be constructed, for example by fitting certain dynamical equations to empirically observed causal regularities (as in population cycles of predator–prey systems in ecology) or by analyzing statistical correlations (as in models of stock-market behaviour). In the economic and social sciences, identifying the relevant parameters and constructing a mathematical model that connects them may often precede theory construction. Frequently, what scientists are interested in are qualitative features, such as the stability or instability of certain systems, and these may be reflected better by a mathematical model than by any available partial evaluation of the underlying theory.

Given this diversity, it would be hopeless to look for shared properties held in common by all mathematical models. Fortunately, there are other ways one can approach the problem. First, the characteristics one is most interested in need not themselves be mathematical properties, but may encompass 'soft' factors such as ease of use, elegance, simplicity and other factors pertaining to the uses to which mathematical models are typically put. Second, it may be possible to identify a subclass of mathematical models—such as the many-body models to be discussed in this paper—which is sufficiently comprehensive to allow for generalizations, but whose members are not too disparate. Finally, it will often be possible to glean additional insight from contrasting mathematical models with other, more general, kinds of models.

On one influential general account, which will prove congenial to the present paper, models are to be regarded as 'mediating instruments' [see (Morrison and Morgan 1999b)]. It is crucial to this view that models are not merely understood as an unavoidable intermediary step in the application of general theories to specific situations. Rather, as 'mediators' between our theories and the world, models inform the interpretation of our theories just as much as they allow for the application of these theories to nature. As Morrison and Morgan are keen to point out, "models are *not* situated in the middle of an hierarchical structure between theory and the world", but operate outside the hierarchical 'theory–world axis' (Morrison and Morgan 1999b, p. 17f.). This can be seen by realizing that models "are made up from a *mixture* of elements, including those from outside the original domain of investigation" (p. 14). It is this partial independence from original theory and data that is required in order to allow models to play an autonomous role in scientific enquiry. In this respect, Margaret Morrison and Mary Morgan argue, scientific models are much like scientific instruments. Indeed, it is part and parcel of this view that model building involves an element of creativity and skill—it is "not only a craft but also an art, and thus not susceptible to rules" (Morrison and Morgan 1999b, p. 12).

A number of case studies have examined specific examples from the natural and social sciences from within this framework. A cross-section of these are collected in (Morrison and Morgan 1999a). The upshot of many of these studies is that "model construction involves a complex activity of integration" (Morrison 1999, p. 44). This integration need not be perfect and, as Daniela Bailer-Jones points out, may involve "a whole range of different means of expression, such as texts, diagrams or mathematical equations" (Bailer-Jones 2003, p. 60). Quite often, the integration cannot be perfect, as certain elements of the model may be incompatible with one another. Even in cases where successful integration of the various elements is possible, the latter can be of very different sorts—they may differ not only in terms of their medium of expression (text, diagram, formula) but also in terms of content: some may consist in mathematical relations, others may draw on analogies; some may reflect actual empirical data, others, perhaps in economics, may embody future target figures (e.g., for inflation).

It is in comparison with this diversity of general aspects of scientific models, I argue, that several characteristic features of *mathematical models* can be singled out. The first of these concerns the medium of expression, which for mathematical models is, naturally, the formal language of mathematics. It would, however, be misguided to simply regard a model as a set of (uninterpreted) mathematical equations, theorems and definitions, as this would deprive models of their empirical relevance: a set of equations cannot properly be said to 'model' anything, neither a specific phenomenon nor a class of phenomena, unless some of the variables are interpreted so as to relate them to observable phenomena. One need not be committed to the view (as Morrison paraphrases Nancy Cartwright's position on the matter) that "fundamental theory represents nothing, [that] there is simply nothing for it to represent since it doesn't describe any real world situations" (Morrison 1998, p. 69), in order to acknowledge that mathematical models cannot merely be uninterpreted mathematical equations if they are to function as mediators of any sort; that is, if they are to *model* a case that, for whatever reason, cannot be calculated or described in terms of theoretical first principles.

The fact that mathematical models, like other kinds of models, require background assumptions and rules of interpretation, of course, does not rule out the fact that in each case there may be a core set of mathematical relationships that model *users* regard as definitive of the mathematical model in question. Indeed, this assumption should be congenial to the proposed analysis of models as mediators, as the mathematical features of a model—where these are not merely 'inherited' from a fundamental theory—may provide it with precisely the autonomy and independence (from theory and data) that the role as mediator requires. This applies especially to the case of many-body models which, as I shall discuss in the Sect. 11.4, are typically the output of what has been called 'mature mathematical formalisms' (in this case, the formalism of second quantization, as adapted to the case of many-body physics).

While it may be true that, as Giere puts it, "[m]uch mathematical modeling proceeds in the absence of general principles to be used in constructing models" (Nersessian and Thagard 1999, p. 52), there are good terminological reasons to speak of a mathematical model *of* a phenomenon (or a class of phenomena) only if the *kinds* of mathematical techniques and concepts employed are in some way sensitive to the *kind* of phenomenon in question. For example, while it may be possible, if only retrospectively, to approximate the stochastic trajectory of a Brownian particle by a highly complex deterministic function, for example a Fourier series of perfectly periodic functions, this would hardly count as a good mathematical model: there is something about the phenomenon, namely its stochasticity, that would not be adequately reflected by a set of deterministic equations; such a set of equations would quite simply not be a mathematical model *of Brownian motion*.[1]

In addition to the requirement that the core mathematical techniques and concepts be sensitive to the kind of phenomenon that is being modelled, there is a further condition regarding what should count as a *mathematical* model. Loosely speaking, the mathematics of the model should do some work in integrating the elements of the 'extended' model, where the term 'extended' refers to the additional

---

[1] There may, of course, be independent reasons why one might represent, say, *a specific trajectory* by a certain set of deterministic equations, or by a (non-mathematical) pictorial representation. However, in such cases, as well as in contexts where the stochasticity of the causal process is irrelevant, one would not be dealing with a model *of Brownian motion*, in the proposed narrower sense of 'mathematical model'.

information needed to apply a bare mathematical structure to individual cases. If, for example, a mathematical model employs the calculus of partial differential equations, then it should also indicate which (classes of) initial and boundary conditions need to be distinguished; likewise, if a mathematical model depends crucially on certain parameters, it should allow for systematic methods of varying, or 'tweaking', those parameters, so that their significance can be studied systematically.[2] This capacity of successful models to integrate different cases, or different aspects of the same case, has occasionally been called 'moulding' (Boumans 1999, p. 90; Bailer-Jones 2003, p. 62):

> Mathematical moulding is shaping the ingredients in such a mathematical form that integration is possible, and contains two dominant elements. The first element is moulding the ingredient of mathematical formalism in such a way that it allows the other elements to be integrated. The second element is calibration, the choice of the parameter values, again for the purpose of integrating all the ingredients. (Boumans 1999, p. 90)

Successful mathematical models, on this account, display a capacity to integrate different elements—some theoretical, others empirical—by deploying an adaptable, yet principled formalism that is mathematically characterizable, (largely) independently of the specifics of the theory and data in the case under consideration.

For the remainder of the present paper, I shall therefore be relying on an understanding of many-body models that recognizes their dual status as models of physical systems (which, importantly, may include purely hypothetical systems) and as mathematical structures. This is in line with the following helpful characterization presented by Sang Wook Yi:

> What I mean by a model in this paper is a mathematical structure of three elements: basic entities (such as 'spins'), the postulated arrangement of the basic entities (say, 'spins are located on the lattice point') and interactions among the basic entities ('spin-magnetic field interactions'). As a rough criterion, we may take a model to be given when we have the Hamiltonian of the model and its implicit descriptions that can motivate various physical interpretations (interpretative models) of the model. (Yi 2002, p. 82)

If this sounds too schematic, or too general, then perhaps a look at some historical examples will make vivid how many-body models have been put to use in condensed matter physics.

11.3 A Brief History of Many-Body Models

In this and the next section, a class of mathematical models will be discussed that was first developed in connection with research on the magnetic properties of solids. The standard way of picturing a solid as a crystal, with the atoms arranged in

---

[2] Systematic 'tweaking', as Martin Krieger observes, "has turned out to be a remarkably effective procedure" (Krieger 1981, p. 428). By varying contributions to the model, e.g., by adding disturbances, one can identify patterns in the response of the model, including regions of stability.

a highly ordered lattice so as to display certain spatial symmetries, and the electrons possibly delocalized, as in a metal, already contains a great deal of assumptions that may or may not be realized in a given physical system. In order to regard this general characterization as a faithful representation of any real physical object, for example of a lump of metal in a given experiment, certain background assumptions have to be in place. For example, it has to be assumed that the piece of metal, which more often than not will display no crystalline structure to the naked eye, really consists of a number of microcrystals, each of which is highly ordered; that the imperfections, which may arise at the boundaries of two adjoining microcrystals or from the admixture of contaminating substances, are negligible; that, for the purpose of the experiment, the description in terms of ions and electrons is exhaustive (for example, that no spontaneous generation of particles occurs, as may happen at high energies).

Picturing a solid as a lattice consisting of ions and electrons is, of course, a rather rudimentary model, as it does not yet tell us anything (except perhaps by analogies we may draw with macroscopic mechanical lattices) about the causal and dynamic features of the system. For this, the acceptance of a physical theory is required—or, in the absence of a theoretical account of the full system, the construction of a mathematical *many-body model*. (Often a physical theory—to the extent that it is accessible by researchers—will include general principles that constrain, but underdetermine, the specifics of a given system). The earliest many-body model of the kind to be discussed in this paper was the *Ising model*, proposed in 1925 by the German physicist Ernst Ising at the suggestion of his then supervisor Wilhelm Lenz. It was published under the modest title 'A Contribution to the Theory of Ferromagnetism' and its conclusions were negative throughout. According to the summary published in that year's volume of *Science Abstracts*, the model is

> an attempt to modify Weiss' theory of ferromagnetism by consideration of the thermal behavior of a linear distribution of elementary magnets which (in opposition to Weiss) have no molecular field but only a non-magnetic action between neighboring elements. It is shown that such a model possesses no ferromagnetic properties, a conclusion extending to a three-dimensional field.[3]

Ising's paper initially did not generate much interest among physicists, as perhaps one would expect of a model that self-confessedly fails to describe the phenomenon for which it was conceived. It was not until the late 1930s that Ising's paper was recognized as displaying a highly complex mathematical behaviour, which, as one contemporary physicist puts it, "continues to provide us with new insights" (Fisher 1983, p. 47).[4]

As a model of ferromagnetic systems the Ising model pursues the idea that a magnet can be thought of as a collection of elementary magnets, whose orientation

---

[3] Quoted in (Hughes 1999, p. 104).

[4] The domain of application has broadened further in recent years. The Ising model is now also used to model networks, spin glasses, population distributions, etc. See, for example, (Matsuda 1981; Ogielski and Morgenstern 1985; Galam 1997).

determines the overall magnetization of the system. If all the elementary magnets are aligned along the same axis, then the system will be perfectly ordered and will display a maximum value of the magnetization. In the simplest one-dimensional case, such a state can be visualized as a chain of 'elementary magnets', all pointing the same way:

$$\cdots \quad \uparrow \quad \uparrow \quad \uparrow \quad \uparrow \quad \uparrow \quad \uparrow \quad \uparrow \quad \uparrow \quad \cdots$$

The alignment of elementary magnets can either be brought about by a strong enough external magnetic field or it can occur spontaneously, as will happen below a critical temperature, when certain substances (such as iron and nickel) undergo a ferromagnetic phase transition. The parameter that characterizes a phase transition, in this case the magnetization M, is also known as the *order parameter* of the transition. The guiding principle behind the theory of phase transitions is that discontinuities in certain thermodynamic quantities can occur spontaneously as a result of the system minimizing other such quantities in order to reach an equilibrium state. Hence, if the interaction between individual elementary magnets i, j, characterized by a constant J_{ij} is such that it favours the parallel alignment of elementary magnets, then one can hope to expect a phase transition below a certain temperature. The energy function of the system as a whole will, therefore, play an important role in the dynamics of the model, and indeed, in the language of mathematical physics, this is what constitutes the many-body model. In the language of 'mathematical moulding', the energy function will be the core element of the many-body model. In the case of the Ising model, this function can be simply expressed as the sum over all interactions of one elementary magnet with all the others (the variable S_i represents the elementary magnet at lattice site i and takes the values $+1$ or -1 depending on the direction in which the elementary magnet points; the minus sign is merely a matter of convention):

$$E = -\sum_{i,j} J_{ij} S_i S_j \, .$$

If one restricts the interaction to nearest neighbours only and assumes that $J_{i,i\pm1} > 0$, then it is obvious that the energy will be minimized when all the elementary magnets point in the same direction, that is when $S_i S_{i+1} = +1$ for all i.

As Ising himself acknowledged, the one-dimensional model fails to predict a spontaneous magnetization, where the latter can simply be defined as the sum over the orientations $(S_i = \pm1)$ of all elementary magnets, in the absence of an external field, divided by their total number:

$$M = \frac{1}{N} \sum_i S_i \, .$$

The reason for the absence of a spontaneous magnetization in the case of the Ising 'chain' lies essentially in the instability, at finite temperatures $(T \neq 0)$, of a presumed ordered state against fluctuations.[5] In the truly one-dimensional case, the chain is infinitely extended $(N \rightarrow \infty)$, and the contribution of an individual elementary magnet to the total system is of only infinitesimal significance. However, one need only introduce one defect—that is, one pair of antiparallel (rather than parallel) elementary magnets—in order to eliminate the assumed magnetization, as the orientations of the elementary magnets on either side of the 'fault line' will cancel out (see the figure below). Given that even the least 'costly' (in terms of energy) fluctuation will destroy the magnetization, the presumed ordered state cannot obtain.

$$\cdots \quad \uparrow \ \uparrow \ \uparrow \ \uparrow \ \uparrow \ \uparrow \ \uparrow \ \downarrow \ \downarrow \ \downarrow \ \downarrow \ \downarrow \ \downarrow \ \downarrow \quad \cdots$$

Whereas Ising's proof of the non-occurrence of a phase transition in one dimension has stood up to scrutiny, Ising's conjecture that the same holds also for the two- and three-dimensional case has since been proven wrong. In 1935, Rudolf Peierls demonstrated that the two-dimensional Ising model exhibits spontaneous magnetization below a critical temperature $T_c > 0$. This marked a turning point in the 'career' of the Ising model as an object of serious study. In what has been described as "a remarkable feat of discrete mathematics" (Hughes 1999, p. 106), Lars Onsager was able to produce an exact solution, at all temperatures, of the two-dimensional version of the Ising model (Onsager 1944). His results concerned not only the existence, or absence, *in general* of a phase transition, but they also delivered a precise value of the critical temperature (at least for the square lattice) and gave a rigorous account of the behaviour of other quantities, such as the specific heat. (See [Brush 1967; Niss 2005] for a more detailed study of the history of the Ising model.)

In summary, the lessons of this brief history of many-body models are as follows. First, it is worth reminding oneself that *as a model of ferromagnetism*, the Ising model was initially considered a failure. At the time Ising proposed his model in 1925, he recognized that its failure lay in not predicting a spontaneous magnetization in one dimension (and, as Ising wrongly conjectured, also in three dimensions). By the time it was recognized, by Peierls and Onsager, that the model *could* explain the occurrence of a phase transition in two (and possibly three) dimensions, however, the theory of ferromagnetism had moved on. For one, Werner Heisenberg, in a paper in 1928, had proposed a quantum theoretical model, essentially by replacing the number-valued variable S_i in the Ising model by operator-valued vectors $\hat{S}_i$. At first glance, this formal change may seem minor, but it indicates a radical departure from the classical assumptions that Ising's model was based on. Where Ising had to postulate the existence of 'elementary magnets', Heisenberg was able to give a physical interpretation in terms of the newly

---

[5] The zero-temperature case $(T = 0)$ is of special significance in a variety of many-body models. However, in order to keep the presentation accessible, $T \neq 0$ will be assumed throughout the following discussion.

discovered spin of atoms and electrons. The departure from classical assumptions also manifests itself mathematically in the use of spin operators, together with their commutation relations (which have no equivalent in classical physics), and this fundamentally changes the algebraic properties of the mathematical core of the model. The novelty of quantum theory, and of Heisenberg's model, however, is only one reason why, despite Peierls and Onsager's seeming vindication, the Ising model did not gain a foothold as a good model of ferromagnetism. For, as Bohr (1911) and van Leeuwen (1921) had rigorously shown, independently of each other, a purely classical system that respects the (classical) laws of electrodynamics, could never display *spontaneous* magnetization (though, of course, it may develop a non-zero magnetization in an external field). Hence, the explanatory power of the Ising model as a model of spontaneous ferromagnetism was doubly compromised: it could not offer an explanation of why there should be 'elementary magnets' in the first place, and it purported to model, using the conceptual repertoire of classical physics, a phenomenon that could be shown to be incompatible with classical physics.[6]

One might question whether at any point in time the Ising model could have been a good model of ferromagnetism. Had Onsager's solution already been published by Ising, could Heisenberg in his 1928 paper still have dismissed Ising's model as "not sufficient to explain ferromagnetism" (Heisenberg 1928)? Hardly, one might argue. But as things stand, this is not what happened. Models are employed in fairly specific contexts, and in the case of mathematical models in particular, the uses to which they are put determine their empirical content. As Bailer-Jones argues, it is "[t]he model users' activity of intending, choosing and deciding [that] accounts for the fact that models, as they are formulated, submit to more than sheer data match" (Bailer-Jones 2003, p. 71). Applying this idea to the Ising model with its varied history, one could perhaps argue that even a model that was initially considered a failure may experience a comeback later, when it is used to model other phenomena or is considered as a testing ground for new theoretical techniques or mathematical concepts—only, of course, that *this* Ising model, now conceived of as an instrument for generating rigorous results and exact solutions for their own sake, would no longer be a model *of ferromagnetism*.

11.4 Constructing Quantum Hamiltonians

Because in the Heisenberg model the hypothetical 'elementary magnets' of the Ising model are replaced by quantum spins and the nature of 'spin' as a non-classical internal degree of freedom is accepted (by fully embracing the algebraic

---

[6] As Martin Niss notes, during the first decade of the study of the Lenz–Ising model "[c]omparisons to experimental results were almost absent", and moreover, its initial "development was not driven by discrepancies between the model and experiments" (Niss 2005, pp. 311–312).

peculiarities of spin operators), this model is a much better candidate for mimicking spontaneous magnetization. Nonetheless, it still represents the spins as rigidly associated with nodes of a lattice in real (geometrical) space. This is plausible for magnetic insulators but not for substances such as iron and nickel where the electrons are mobile and can, for example, sustain an electric current. However, one can define a concept of *pseudo-spins*, which retains the idea that spins can interact directly, even when it is clear that, in a metal with delocalized electrons, all spin–spin interactions must eventually be mediated by the entities that are *in fact* the spin carriers—that is, electrons. 'Pseudo-spins' were first constructed mathematically via the so-called electron number operators. However, this mathematical mapping of different kinds of operators onto each other does not yet result in a self-contained, let alone intuitive, many-body model for systems with delocalized electrons. This is precisely the situation a model builder finds herself in when she sets out to construct a model for a specific phenomenon, or class of phenomena, such as the occurrence of spontaneous magnetization in a number of physical materials. Since 'fundamental theory' allows for almost limitless possible scenarios, the challenge lies in constructing a model that is *interpretable* by the standards of the target phenomenon. Questions of empirical accuracy—which, after all, cannot be known in advance—are secondary during the phase of model construction. If a model is indeed an instrument of inquiry and, as some claim, if it is "inherently intended for specific phenomena" (Suárez 1999, p. 75), then at the very least there must be a way of interpreting some (presumably the most salient) elements of the model as representing a feature of the phenomenon or system under consideration. This demand has direct implications for how models are constructed. If models do indeed aim at representing their target system, then success in constructing models will be judged by their power to represent. Thus, among proponents of the models-as-mediators view, it is a widely held view that "[t]he proof or legitimacy of the representation [by a model] arises as a result of the model's performance in experimental, engineering and other kinds of interventionist contexts" (Morrison 1998, p. 81). One would expect, then, that the process of model construction should primarily be driven by a concern for whether or not its product—the models—are empirically successful.

By contrast, I want to suggest that the case of many-body models is a paradigmatic example of a process of model construction that neither regards models as mere limiting cases of 'fundamental theory' nor appeals to empirical success as a guide to (or, indeed, the goal of) model construction. Instead, it involves the interplay of two rather different strategies, which I shall refer to as the *first-principles* (or ab initio) approach, on the one hand, and the *formalism-driven* approach on the other. Whereas the expression 'formalism-driven' is my coinage, the first pair of expressions—'first principles' and '*ab initio*'—reflects standard usage in theoretical condensed matter physics, where it is used in contradistinction to so-called 'phenomenological' approaches which aim to develop models by interpolating between specific empirical observations:

The *first principles* approach to condensed matter theory is entirely different from this. It starts from what we know about all condensed matter systems—that they are made of atoms, which in turn are made of a positively charged nucleus, and a number of negatively charged electrons. The interactions between atoms, such as chemical and molecular bonding, are determined by the interactions of their constituent electrons and nuclei. All of the physics of condensed matter systems arises ultimately from these basic interactions. If we can model these interactions accurately, then all of the complex physical phenomena that arise from them should emerge naturally in our calculations (Gibson 2006, p. 2).

A clear, but overambitious, example of a first-principles approach would be the attempt to calculate the full set of $\sim 10^{23}$ coupled Schrödinger equations, one for each of the $\sim 10^{23}$ nodes in the crystal lattice. For obvious reasons, solving such a complex system of equations is not a feasible undertaking—indeed, it would merely restate the problem in the terms of fundamental theory, the complexity of which prompted the introduction of (reduced) models in the first place. But less ambitious, and hence more tractable, first-principles approaches exist. Thus, instead of taking the full system—the extended solid-state crystal—as one's starting point, one may instead begin from the smallest 'building block' of the extended crystal, by considering the minimal theory of two atoms that are gradually moved together to form a pair of neighbouring atoms in the crystal. One can think of this way of constructing models as involving a thought experiment regarding how a many-body system 'condenses' from a collection of isolated particles. Such an approach, although it does not start from the 'full' theory of all $\sim 10^{23}$ particles, remains firmly rooted in 'first principles', in that the thought experiment involving the two 'neighbouring' atoms approaching one another is being calculated using the full theoretical apparatus (in this case, the theoretical framework of non-relativistic quantum mechanics).[7] This is the 'derivation' of many-body models that is usually given in textbooks of many-body theory [e.g., (Nozières 1963)], often with some degree of pedagogical hindsight. However, while such a derivation makes vivid which kinds of effects—e.g., single-particle kinetic energy, particle–particle Coulomb repulsion, and genuine quantum exchange interactions between correlated particles—may be expected to become relevant, it typically remains incomplete as a model of the extended many-body system: what is being considered is only the smallest 'building block', and a further constructive move is required to generate a many-body model of the full crystal.

This is where the second kind of procedure in model construction—what I shall call the *formalism-driven* approach—needs to be highlighted. This approach, in my view, is far more ubiquitous than is commonly acknowledged, and it sheds light on the interplay between mathematical formalism and rigor on the one hand, and the interpretation of models and the assessment of their validity on the other. In particular, it also reinforces the observation that many-body models enjoy a considerable degree of independence from specific experimental (or other interventionist)

---

[7] Needless to say, a considerable number of background assumptions are necessary in order to identify which unit is indeed the smallest one that still captures the basic mechanisms that determine the behaviour of the extended system.

contexts, and even from quantitative standards of accuracy. On the account I am proposing, a "mature mathematical formalism" is "a system of rules and conventions that deploys (and often adds to) the symbolic language of mathematics; it typically encompasses locally applicable rules for the manipulation of its notation, where these rules are derived from, or otherwise systematically connected to, certain theoretical or methodological commitments" (Gelfert 2011, p. 272). In order to understand how the formalism-driven strategy in model construction works, let us return to the case under consideration, namely ferromagnetic systems with itinerant electrons.

How is one to model the itinerant nature of conduction electrons in such metals as cobalt, nickel, and iron? The formalism of so-called creation and annihilation operators, $\hat{a}^{\dagger}_{i,\sigma}$ and $\hat{a}_{i,\sigma}$, allows one to describe the dynamics of electrons in a crystal. Since electrons cannot simply be annihilated completely or created *ex nihilo* (at least not by the mechanisms that govern the dynamics in a solid at room temperature), an annihilation operator acting at one lattice site must always be matched by a creation operator acting at another lattice site. But this is precisely what describes itinerant behaviour of electrons in the first place. Hence, the formalism of second quantization, in conjunction with the basic assumption of preservation of particle number, already suggests how to model the kinetic behaviour of itinerant electrons, namely through the following contribution to the Hamiltonian:

$$\hat{H}_{\text{kin}} = \sum_{ij\sigma} T_{ij}\hat{a}^{\dagger}_{i,\sigma}\hat{a}_{j,\sigma} \ .$$

When the operator product $\hat{a}^{\dagger}_{i,\sigma}\hat{a}_{j,\sigma}$ acts on a quantum state, it first[8] annihilates an electron of spin σ at lattice site j (provided such an electron happens to be associated with that lattice site) and then creates an electron of spin σ at another site i. Because electrons are indistinguishable, it appears, from within the formalism, as if an electron of spin σ had simply moved from j to i. The parameters T_{ij}, which determine the probability of occurrence of such electron 'hopping' from one place to another, are known as *hopping integrals*. In cobalt, nickel, and iron, the electrons are still comparatively tightly bound to their associated ions, so hopping to distant lattice sites will be rare. This is incorporated into the model for the kinetic behaviour of the electrons by including in the model the assumption that hopping only occurs between nearest neighbours.

Hopping is not the only phenomenon that a model for itinerant electrons should reflect. One must also consider the Coulomb force between electrons—that is, the fact that two negatively charged entities will experience electrostatic repulsion. Once again, the formalism of second quantization suggests a straightforward way to

---

[8] Operators should be read from right to left, so if an operator product like $\hat{a}^{\dagger}_{i,\sigma}\hat{a}_{j,\sigma}|\Psi\rangle$ acts on a quantum state, the operator $\hat{a}_{j,\sigma}$ directly in front of $|\Psi\rangle$ acts first, followed by $\hat{a}^{\dagger}_{i,\sigma}$. Because operators do not always commute, the order of operation is important.

account for the Coulomb contribution to the Hamiltonian. Since the Coulomb force will be greatest for electrons at the same lattice site (which must then have different spins, due to the Pauli exclusion principle), the dominant term will be

$$\hat{H}_{\text{Coulomb}} = \sum_{i\sigma} \frac{U}{2} \hat{n}_{i,\sigma} \hat{n}_{i,-\sigma} .$$

The sum of these two terms—the hopping term (roughly, representing movement of electrons throughout the lattice) and the Coulomb term (the potential energy due to electrostatic repulsion)—already constitutes the Hubbard model:

$$\hat{H}_{\text{Hubbard}} = \hat{H}_{\text{kin}} + \hat{H}_{\text{Coulomb}}$$

$$= \sum_{ij\sigma} T_{ij} \hat{a}_{i,\sigma}^{\dagger} \hat{a}_{j,\sigma} + \sum_{i\sigma} \frac{U}{2} \hat{n}_{i,\sigma} \hat{n}_{i,-\sigma}.$$

Note that, unlike the first-principles approach, the formalism-based approach to model construction does not begin with a description of the physical situation in terms of fundamental theory, either in the form of the 'full' set of $\sim 10^{23}$ coupled Schrödinger equations, or via the thought experiment of neighbouring atoms gradually approaching each other so as to form the elementary 'building block' of an extended crystal lattice. Instead, it models the presumed microscopic processes (such as hopping and Coulomb interaction) *separately*, adding up the resulting components and, in doing so, constructing a many-body model 'from scratch', as it were, without any implied suggestion that the Hamiltonian so derived is the result of approximating the full situation as described by fundamental theory. Interestingly, Nancy Cartwright argues against what she calls "a mistaken reification of the separate terms which compose the Hamiltonians we use in modelling real systems" (Cartwright 1999, p. 261). Although Cartwright grants that, on occasion, such terms "represent separately what it might be reasonable to think of as distinct physical mechanisms", she insists that "the break into separable pieces is purely conceptual" (ibid.) and that what is needed are "independent ways of identifying the representation as correct" (Cartwright 1999, p. 262). Cartwright's critique of formalism-based model construction must be understood against the backdrop of her emphasis on phenomenological approaches, which she regards as the only way "to link the models to the world" (ibid.). To be sure, the formalism-driven approach often proceeds in disregard of specific empirical phenomena and in this respect might be considered as remote from Cartwright's preferred level of description—the world of physical phenomena—as the more 'first-principles'-based approaches. But it would be hasty to reject the formalism-driven approach for this reason alone, just as it would be hasty to consider it simply an extension of 'fundamental theory'. It is certainly true that the formalism-driven approach is not theory-free. But much of the fundamental theory is hidden in the formalism—the formalism, I have argued elsewhere, may be said to 'enshrine' various theoretical, ontological, and methodological commitments and assumptions [see (Gelfert 2011)]. Consider, for example, how the construction of the kinetic part of the model proceeded from purely heuristic considerations of how itinerant motion in a discrete lattice could be pictured intuitively in terms of the annihilation of an electron at one place in the lattice and its subsequent creation at another. The hopping integrals T_{ij} were even introduced as mere parameters, when, on the first-principles approach, they ought to be interpreted as matrix elements, which contain the bulk of what quantum theory can tell us about the probability amplitude of such events. Finally, the Coulomb term was constructed almost entirely by analogy with the classical case, except for the reference to the Pauli principle. (Then again, the Pauli principle itself is what makes the formalism of second quantization and of creation/annihilation operators work in the first place—a fact that the present formalism-driven derivation did not for a moment have to reflect upon.[9]) Rather than thinking of the formalism-based approach as drawing a veil over the world of physical phenomena, shrouding them in a cocoon of symbolic systems, one should think of formalisms such as the many-body operators discussed above as playing an *enabling* role: not only do they allow the model builder to represent selected aspects of complex systems, but in addition one finds that "in many cases, it is *because* Hamiltonian parts can be interpreted literally, drawing on the resources furnished by fundamental theory as well as by (interpreted) domain-specific mathematical formalisms, that they generate *understanding*" (Gelfert 2013, p. 264; see also Sect. 11.5.3). While the formalism-based approach is not unique in its ability to model selected aspects of complex systems (in particular, different co-existing 'elementary' processes), it does so with an especially high degree of economy, thereby allowing the well-versed user of a many-body model to develop a 'feel' for the model and to probe its properties with little explicit theoretical mediation.

11.5 Many-Body Models as Mediators and Contributors

Earlier, I argued that mathematical models should be sensitive to the phenomena they are intended to model. As argued in the last section (11.4), the existence of a mature formalism—such as second quantization with its rules for employing creation and annihilation operators—can guarantee certain kinds of sensitivity, for example the conformity of many-body models to certain basic theoretical commitments (such as the Pauli principle). At the same time, however, the formalism frees the model from some empirical constraints: by treating the hopping integrals as parameters that can be chosen largely arbitrarily (except perhaps for certain symmetry requirements), it prevents the relationship of sensitivity between model and empirical data from turning into a relationship of subjugation of the model by the data.

---

[9] For a discussion of the formalism of creation and annihilation operators as a 'mature mathematical formalism', see (Gelfert 2011, pp. 281—282).

On a standard interpretation, applying models to specific physical systems is a two-step process. First, a 'reduced' mathematical model is derived from fundamental theory (a simplistic view that has already been criticized earlier); second, approximative techniques of numerical and analytical evaluation must be employed to calculate physical observables from the model, again at the expense of the mathematical complexity of the (still not exactly solvable) Hamiltonian. However, this way of speaking of two successive steps of approximation puts undue emphasis on the loss of accuracy involved in the process. For, it is not clear how lamentable this 'loss' really is, given the unavailability of an exact solution to the full problem. Crucially, such a view also overlooks the fact that the model itself contributes new elements to the theoretical description of the physical system, or class of systems, under consideration—elements, which are not themselves part of the fundamental theory (or, as it were, cannot be 'read off' from it) but which may take on an interpretative or otherwise explanatorily valuable role.

Contributions of this sort, originating from the model rather than from either fundamental theory or empirical data, do, however, considerably inform the way physicists think about a class of systems and frequently suggest new lines of research. Consider the limitation to two sets of parameters U, $\{T_{ij}\}$ in the case of the Hubbard model. Assuming a cubic lattice and nearest-neighbour interaction, the interaction T_{ij} between different lattice sites will either be zero or have the same fixed value t. Hence, the quotient U/t reflects the relative strength of the interaction *between* electrons (as compared with their individual kinetic movement), and *within the model* it is a unique and exact measure of this important aspect of the dynamics of electron behaviour in a solid. The individual quantities U and t are thus seen to be no mere parameters, but are linked, through the model, in a meaningful way, which imposes constraints on which precise values are, or aren't, plausible. Not only does this restrict the freedom one enjoys in arbitrarily choosing U and t to fit the model to the empirical data, but it also imposes constraints on structural modifications of the model. For example, an attempt to make the model more accurate by adding new (higher-order) terms to the model (perhaps accounting for higher-order interactions of strengths $V, W, X, Y, Z < U, t$), may be counterproductive, as it may be more useful, for explanatory purposes, to have *one* measure of the relative strength of the electron–electron interaction (namely, U/t) rather than a whole set $\{U/t, V/t, W/t, \ldots\}$. To the extent that the model's purpose is explanatory and not merely predictive, a gain in numerical accuracy may not be desirable if it involves replacing an intuitively meaningful quantity with a set of parameters that lack a straightforward interpretation. Fitting a model to the data does not by itself make the model any more convincing.

The 'active' contribution of the model, that is, the fact that it contributes *new elements* rather than merely *integrating* theoretical and experimental (as well as further, external) elements, is not only relevant to interpretative issues, but also has direct consequences for assessing the techniques used to evaluate the model and to calculate, either numerically or analytically, observable quantities from it.

11.5.1 Rigorous Results and Relations

One particularly salient class of novel contributions that many-body models make
to the process of inquiry in condensed matter physics is known as *rigorous results*.
The expression 'rigorous results', which is not without its problems, has become a
standard expression in theoretical physics, especially among practitioners of statistical and many-body physics [see, for example, (Baxter 1982)]. It therefore calls
for some clarification. What makes a result 'rigorous' is not the qualitative or
numerical accuracy of a particular prediction of the theory or model. In fact, the
kind of 'result' in question will often have no immediate connection with the
empirical phenomenon (or class of phenomena) a model or theory is supposed to
explain. Rather, it concerns an exact *mathematical* relationship between certain
mathematical variables, or certain structural components, of the mathematical
model, which may or may not reflect an empirical feature of the system that is being
modelled. One, perhaps crude, way of thinking about rigorous results would be to
regard them as mathematical theorems that are provable from within the model or
theory under consideration.[10] Much as Pythagoras' theorem $a^2 + b^2 = c^2$ is not
merely true of a particular set of parameters, e.g., $\{a, b, c\} = \{3, 4, 5\}$, but holds for
all rectangular triangles, so a rigorous result in the context of a mathematical model
holds for a whole class of cases rather than for particular parameter values. Yet,
importantly, rigorous results are true only *of a model* (or a class of models) as
defined by a specific Hamiltonian; unlike, say, certain symmetry or conservation
principles, they do not follow directly from fundamental theory.

   An important use of rigorous results and relations is as 'benchmarks' for the
numerical and analytical techniques for calculating observable quantities from the
model.[11] After all, an evaluative technique that claims to be true to the model
should preserve its main features, and rigorous results often take the form either of
exact relations holding between two or more quantities, or of lower and upper
bounds on certain observables. If, for example, the order parameter in question is
the magnetization, then rigorous results—within a given model—may obtain,
dictating the maximum (or minimum) value of the magnetization or the magnetic
susceptibility. These may then be compared with results derived numerically or by
other approximative methods of evaluation.

---

[10] The notions of 'theorem' and 'rigorous result' are frequently used interchangeably in scientific
texts, especially in theoretical works such as (Griffiths 1972).

[11] This is noted in passing, though not elaborated on, by R.I.G. Hughes in his case study of one of
the first computer simulations of the Ising model (Hughes 1999, p. 123): "In this way the veri-
similitude of the simulation could be checked by comparing the performance of the machine
against the exactly known behaviour of the Ising model."

11.5.2 Cross-Model Support

Rigorous results may also connect different models in unexpected ways, thereby allowing for cross-checks between methods that were originally intended for different domains. Such connections can neither be readily deduced from fundamental theory, since the rigorous results do not hold *generally* but only between different (groups of) models; nor can they justifiably be inferred from empirical data, since the physical systems corresponding to the two groups of mathematical many-body models may be radically different. As an example, consider again the Hubbard model. It can been shown rigorously [see, for example, (Gebhard 1997)] that, at half filling (that is, when half of the quantum states in the conduction band are occupied) and in the strong-coupling interaction limit $U/t \to \infty$, the Hubbard model can be mapped onto the spin-1/2 antiferromagnetic Heisenberg model (essentially in the form described earlier, with $J_{ij} = 4t^2/U$). Under the specified conditions, the two models are isomorphic and display the same mathematical behaviour. Of course, the Hubbard model with *infinitely* strong electron–electron interaction ($U/t \to \infty$) cannot claim to describe an actual physical system, where the interaction is necessarily finite, but to the extent that various mathematical and numerical techniques can nonetheless be applied in the strong-coupling limit, comparison with the numerically and analytically more accessible antiferromagnetic Heisenberg model provides a test also for the adequacy of the Hubbard model.

Rigorous relations between different many-body models do not only provide fertile ground for testing mathematical and numerical techniques, and for the 'exploration' (in the sense discussed in the next subsection) of models more generally. They can also give rise to a transfer of empirical warrant across models that were intended to describe very different physical systems. The mapping, in the strong-coupling limit ($U/t \to \infty$), of the Hubbard model onto the spin-1/2 antiferromagnetic Heisenberg model is one such example. For the latter—the antiferromagnetic Heisenberg model—has long been known as an empirically successful "'standard model' for the description of magnetic insulators" (Gebhard 1997, p. 75), yet the Hubbard model at low coupling ($U/t = 0$, indicating zero electron–electron interaction) reduces to an ideal Fermi electron gas—a perfect conductor. It has therefore been suggested that, for some finite value between $U/t = 0$ and $U/t \to \infty$, the Hubbard model must describe a system that undergoes a transition from conductor to insulator. Such transitions, for varying strengths of electron–electron interaction, have indeed been observed in physical systems and are known as Mott insulators. Thanks to the existence of a rigorous relation between the two models, initial empirical support for the *Heisenberg* model as a model of a magnetic insulator thus translates into support for a new—and originally unintended—representational use of the *Hubbard* model, namely as a model for Mott insulators. In other words, "empirical warrant first flows from one model to another, in virtue of their standing in an appropriate mathematically rigorous relation" (Gelfert 2009, p. 516), from which one may then gain new insights into the empirical adequacy of the model.[12] As this example illustrates, rigorous results neither borrow their authority from fundamental theory nor do they need to prove their mettle in experimental contexts; instead, they are genuine contributions of the models themselves, and it is through them that models—at least those of the kind discussed in this—have 'a life of their own'.

11.5.3 Model-Based Understanding

The existence of rigorous results and relations, and of individual cases of cross-model support between many-body models of quite different origins, may perhaps seem too singular. Can any general lessons be inferred from them regarding the broader character of many-body models? I wish to suggest that both classes of cases sit well with general aspects of many-body models and their construction, especially when viewed from the angle of the formalism-based approach. By reconceptualizing many-body models as outputs of a mature mathematical formalism—rather than conceiving of them either as approximations of the 'full' (but intractable) theoretical description or as interpolating between specific empirical phenomena—the formalism-based approach allows for a considerable degree of flexibility and exploration, which in turn generates *understanding*. For example, one may construct a many-body model (which may even be formulated in arbitrary spatial dimensions) by imagining a crystal lattice of a certain geometry, with well-formed (by the lights of the many-body formalism) mathematical expressions associated with each lattice point, and adding the latter up to give the desired 'Hamiltonian': "Whether or not this 'Hamiltonian' is indeed the Hamiltonian of a real physical system, or an approximation of it, is not a consideration that enters at this stage of model construction." (Gelfert 2013, p. 262) The phenomenological approach advocated by Cartwright might lament this as creating an undue degree of detachment from the world of empirical phenomena, but what is gained in the process is the potential for exploratory uses of models. As Yi puts it:

> One of the major purposes of this 'exploration' is to identify what the true features of the model are; in other words, what the model can do with and without additional assumptions that are not a part of the original structure of the model. (Yi 2002, p. 87)

Such exploration of the intrinsic features of a model "helps us shape our physical intuitions about the model", even before these intuitions become, as Yi puts it, 'canonical' through "successful application of the model in explaining a phenomenon" (ibid.).

---

[12] This case of cross-model support between many-body models that were originally motivated by very different concerns is discussed in detail in (Gelfert 2009).

Exploratory uses of models feed directly into model-based understanding, yet they do so in a way that is orthogonal to the phenomenological approach and its emphasis on interpolation between observed physical phenomena. As I have argued elsewhere, microscopic many-body models "are often deployed in order to account for poorly understood phenomena (such as specific phase transitions); a premature focus on empirical success (e.g., the exact value of the transition temperature) might lead one to add unnecessary detail to a model before one has developed a sufficient understanding of which microscopic processes influence the macroscopically observable variable" (Gelfert 2013, p. 264). A similar observation is made by those who argue for the significance of minimal models. Thus Robert Batterman argues (quoting a condensed matter theorist, Nigel Goldenfeld):

> On this view, what one would like is a good minimal model—a model "which most economically caricatures the essential physics" (Goldenfeld 1992, p. 33). The adding of details with the goal of 'improving' the minimal model is self-defeating—such improvement is illusory. (Batterman 2002, p. 22)

The formalism-based approach thus differs from the phenomenological approach in two important ways. First, it conceives of model construction as a constructive and exploratory process, rather than as one that is driven by tailoring a model to specific empirical phenomena. This is aptly reflected by Yi in his account of model-based understanding, which posits two stages:

> (1) understanding of the model under consideration, and this involves, among other things, exploring its potential explanatory power using various mathematical techniques, figuring out various plausible physical mechanisms for it and cultivating our physical intuition about the model; (2) matching the phenomenon with a well-motivated interpretative model of the model (Yi 2002, pp. 89–90)

Second, the two approaches differ in the relative weight they accord to empirical adequacy and model-based understanding as measures of the performance of a model. In the formalism-based approach, empirical adequacy is thought of as a 'bonus'—in the sense that "model-based understanding does not necessarily presuppose empirical adequacy" (Yi 2002, p. 85). Such model-based understanding need not be restricted to purely internal considerations, such as structural features of the model, but may also extend to general questions about the world, especially where these take the form of 'how-possibly' questions. For example, in the many-body models under discussion, an important driver of model construction has been the question of how there could possibly arise *any* magnetic phase transition (given the Bohr–van Leeuwen prohibition on spontaneous magnetization in classical systems; see Sect. 11.3)—regardless of any actual, empirically observed magnetic systems. By contrast, the phenomenological approach is willing to trade in understanding of the inner workings of a model for specific empirical success. As Cartwright puts it, "[a] Hamiltonian can be admissible under a model—and indeed under a model that gives good predictions—without being explanatory if the model itself does not purport to pick out basic explanatory mechanisms" (Cartwright 1999, p. 271).

As an illustration of how the formalism-based approach and the phenomenological approach pull in different directions, consider which contributions to a many-body model (that is, additive terms in a Hamiltonian) each approach deems admissible. According to Cartwright, only those terms are admissible that are based on 'basic interpretative models' that have been studied independently and are well-understood, both on theoretical grounds and in *other* empirical contexts; these are the textbook examples of the central potential, scattering, the Coulomb interaction, the harmonic oscillator, and kinetic energy (Cartwright 1999, p. 264). What licenses their use—and, in turn, excludes other (more 'arbitrary' or 'formal') contributions to the Hamiltonian—is the existence of 'bridge principles' which "attach physics concepts to the world" (Cartwright 1999, p. 255). Indeed, Cartwright goes so far as to assert that quantum theory "applies exactly as far as its interpretative models can stretch": only those situations that are captured adequately by the half-dozen or so textbook examples of interpretative models "fall within the scope of the theory" (Cartwright 1999, p. 265). By contrast, the formalism-based approach tells a very different story. As long as one 'plays by the rules' of the formalism—which now *enshrines* theoretical constraints, without the need to make them explicit even to the experienced user—any newly constructed Hamiltonian terms are admissible in principle. And, indeed, in our earlier discussion of how to construct Quantum Hamiltonians we already encountered a contribution to the Hamiltonian—the *hopping* term—which was not inspired by the limited number of stock examples allowed on the phenomenological approach, but instead resulted from a creative application of the formalism-based rules for the 'creation' and 'annihilation' of particles at distinct lattice sites. By freeing model construction from the overemphasis on empirical adequacy, the formalism-based approach not only allows for a more flexible way of modelling specific processes that are thought to contribute to the overall behaviour of a complex system, but gives modellers the theoretical tools to sharpen their understanding of the diverse interactions that together make up the behaviour of many-body systems.

11.6 Between Rigor and Reality: Appraising Many-Body Models

Traditionally, models have been construed as being located at a definite point on the 'theory–world axis' (Morrison and Morgan 1999b, p. 18). Unless their role was seen as merely heuristic, models were to be judged by how well they fit with the fundamental theory and the data, or, more specifically, how well they explain the data by the standards of the fundamental theory. Ideally, a model should display a tight fit with both the theory and the empirical data or phenomena. As Tarja Knuuttila has pointed out, large parts of contemporary philosophy of science continue to focus on "the *model–target* dyad as a basic unit of analysis concerning models and their epistemic values" (Knuuttila 2010, p. 142). The proposed alternative view of models as mediators presents a powerful challenge to the traditional picture. It takes due account of the fact that, certainly from an epistemic point of view, theories can only ever be partial descriptions of what the world is like. What is called for is an account of models that imbues them with the kind of autonomy that does not require a close fit with fundamental theory, but nevertheless enables us to explain and understand physical phenomena where no governing fundamental theory has been identified. On this view, any account of real processes and phenomena also depends on factors that are extraneous to the fundamental theory, and those who deny this, are "interested in a world that is not our world, not the world of appearances but rather a purer, more orderly world, a world which is thought to be represented 'directly' by the theory's equations" (Cartwright 1999, p. 189).

The mediator view of models acknowledges from the start that "it is because [models] are made up from a *mixture* of elements, including those from outside the original domain of investigation, that they maintain [their] partially independent status" (Morrison and Morgan 1999b, p. 14). This is what makes them *mediators* in the first place:

> Because models typically include other elements, and model building proceeds in part independently of theory and data, we construe models as being outside the theory–world axis. It is this feature which enables them to mediate effectively between the two. (Morrison and Morgan 1999b, p. 17f.)

Note that this is essentially a claim about the construction of models, their motivation and etiology. Once a model has been arrived at, however, it is its empirical success in specific interventionist contexts which is the sole arbiter of its validity. This follows naturally from a central tenet of the mediator view, namely that models are closer to instruments than to theories and, hence, warranted by their instrumental success in specific empirical contexts.[13] That models are to be assessed by their specificity to empirically observed phenomena, rather than by, say, theoretical considerations or mathematical properties intrinsic to the models themselves, appears to be a widely held view among proponents of the models-as-mediators view. As Mauricio Suárez argues, models "are inherently intended for specific phenomena" (Suárez 1999, p. 75), and Margaret Morrison writes (Morrison 1998, p. 81): "The proof or legitimacy of the representation arises as a result of the model's performance in experimental, engineering and other kinds of interventionist contexts—nothing more can be said!" It appears then that, whilst the mediator view of models has 'liberated' models from the grip of theory, by stressing their capacity to integrate disparate elements, it has retained, or even strengthened, the close link between models and empirical phenomena.

---

[13] As Cartwright argues, it is for this reason that warrant to believe in predictions must be established case by case on the basis of models. She criticizes the 'vending-machine view', in which "[t]he question of transfer of warrant from the evidence to the predictions is a short one since it collapses to the question of transfer of warrant from the evidence to the theory". This, Cartwright writes, "is not true to the kind of effort hat we know it takes in physics to get from theories to models that predict what reliably happens"; hence, "[w]e are in need of a much more textured, and I am afraid much more laborious view" regarding the claims and predictions of science (Cartwright 1999, p. 185).

Yet, on a descriptive level, it is by no means clear that, for example in the case of the Hubbard model, the main activity of researchers is to assess the model's performance in experimental or other kinds of interventionist contexts. A large amount of work, for example, goes into calibrating and balancing different methods of numerical evaluation and mathematical analysis. That is, the calibration takes place not between model and empirical data, but between different methods of approximation, irrespective of their *empirical* accuracy. Even in cases, where 'quasi-exact' numerical results are obtainable for physical observables (for example via quantum Monte Carlo calculations), these will often be compared not to experimental data but instead to other predictions derived at by other approximative methods. It is not uncommon to come across whole papers on, say, the problem of 'magnetism in the Hubbard model', that do not contain a single reference to empirical data. (As an example, see (Tusch et al. 1996)). Rather than adjust the parameters of the model to see whether the empirical behaviour of a specific physical system can be modelled accurately, the parameters will be held fixed to allow for better comparison of the different approximative techniques with one another, often singling out one set of results (e.g., those calculated by Monte Carlo simulations) as authoritative.

One might object that a good deal of preliminary testing and cross-checking of one's methods of evaluation has to happen before the model predictions can be compared with empirical data, but that nonetheless the latter is the ultimate goal. While there may be some truth to this objection, it should be noted that in many cases this activity of cross-checking and 'bench-marking' is what drives research and makes up the better part of it. It appears that at the very least this calls for an acknowledgment that some of the most heavily researched models typically are not being assessed by their performance in experimental, engineering, and other kinds of interventionist contexts. In part, this is due to many models not being intended for *specific* phenomena, but for a range of physical systems. This is true of the Hubbard model, which is studied in connection with an array of quite diverse physical phenomena, including spontaneous magnetism, electronic properties, high-temperature superconductivity, metal–insulator transitions, and others, and it is particularly obvious in the case of the Ising model, which, even though it has been discredited as an accurate model of magnetism, continues to be applied to problems ranging from soft condensed matter physics to theoretical biology. In some areas of research, models are not even intended, in the long-term, to reflect, or be 'customizable' to, the details of a specific physical system. For example, as R.I.G. Hughes argues, when it comes to critical phenomena "a good model acts as an exemplar of a universality class, rather than as a faithful representation of any one of its members" (Hughes 1999, p. 115).

The reasons why many-body models can take on roles beyond those defined by performance in empirical and interventionist contexts are identical to those that explain their capacity to 'survive' empirical refutation *in a specific context* (as was the case with the Ising model). As I have argued in this paper, they are two-fold. First, models often *actively contribute* new elements, and this introduces cohesion and flexibility. One conspicuous class of such contributions (see Sect. 11.5.1 above) are the rigorous results and relations that hold for a variety of many-body models, without being entailed either by the fundamental theory or the empirical data. It is such rigorous results, I submit, which guide much of the research by providing important 'benchmarks' for the application of numerical and analytical methods. Rigorous results need not have an obvious empirical interpretation in order to guide the search for better techniques of evaluation or analysis. This is frequently overlooked when philosophers of science discuss the role of many-body models. Cartwright, for example, writes:

> When the Hamiltonians do not piggy-back on the specific concrete features of the model—that is, when there is no bridge principle that licenses their application to the situation described in the model—then their introduction is ad hoc and the power of the derived prediction to confirm the theory is much reduced. (Cartwright 1999, p. 195)

It is certainly true that many-body Hamiltonians that do 'piggy-back' on concrete features of the model frequently fare better than more abstract representations—if only because physicists may find the former more 'intuitive' and easier to handle than the latter. But it is questionable whether the absence of 'specific concrete features', which would pick out a specific empirical situation, is enough to render such Hamiltonians ad hoc. For there typically exist additional constraints, in the form of rigorous results and relations, that do constrain the choice of the Hamiltonian, and these may hold for a quite general class of models, irrespective of the specific concrete features of a given empirical case. In particular, the process of 'benchmarking' across models on the basis of such rigorous results and relations is not merely another form of 'moulding' a mathematical model to concrete empirical situations; rather, it fulfills a normative function by generating cross-model cohesion.

The second main reason why the role of many-body models in condensed matter physics is not exhausted by their empirical success lies in their ability to confer insight and understanding regarding the likely microscopic processes underlying macroscopic phenomena, even in the absence of a fully developed theory. As discussed in Sect. 11.5.3, this is directly related to the exploratory use of many-body models—and this in turn is made possible by the formalism-based mode of model-building, which allows for the 'piece-meal' construction of many-body Hamiltonians. Especially in the case of physical systems that are marked by complexity and strong correlations among their constituents, what is aimed for is a model which, in Goldenfeld's apt formulation, "most economically caricatures the essential physics" (Goldenfeld 1992, p. 33).

Given my earlier endorsement of the view that models need to be liberated from the grip of self-proclaimed 'fundamental theories', one might worry that further liberating them of the burden of empirical success leads to an evaporation of whatever warrant models previously had. This is indeed a legitimate worry, and it is one that is shared by many scientists working on just those models. If all there is to a model is a set of mathematical relations together with a set of background assumptions, how can we expect the model to tell us anything about the world? There are several points in reply to this challenge. First, while it is true that many of the rigorous relations do not easily lend themselves to an empirical interpretation, there are, of course, still many quantities (such as the order parameter, temperature, etc.) that have a straightforward empirical meaning. Where the model does make predictions about certain empirically significant observables, these predictions will often be an important (though not the only) measure of the model's significance.[14] Second, models can mutually support each other. As the example of the mapping of the strong-coupling Hubbard model at half-filling onto the Heisenberg model showed, rigorous results and relations can connect different models in unexpected ways. This allows for some degree of transfer of warrant from one model to the other. Note that this transfer of warrant does not involve any appeal to fundamental theory, but takes place 'horizontally' at the level of models.[15] Third, in many cases a model can be constructed in several different ways, which may bring out the connection with both theory and phenomenon in various ways. The first-principles derivation of the Hubbard model is one such example. It provides a meaningful interpretation of the otherwise merely parameter-like quantities T_{ij} (namely, as matrix elements that describe the probability of the associated hopping processes). While this interpretation requires some appeal to theory, it does not require an appeal to the full theoretical problem—that is, the full problem of 10^{23} particles each described by its 'fundamental' Schrödinger equation. A similar point can even be made for the formalism-driven approach. There, too, model construction does not operate in a conceptual vacuum, but makes use of general procedures, which range from the highly abstract (e.g., the formalism of second quantization) to the largely intuitive considerations that go into the selection of elementary processes judged to be relevant.

By recognizing that models can be liberated both from the hegemony of fundamental theory and from the burden of empirical performance in every specific concrete case, I believe one can appreciate the role of models in science in a new light. For one, models are as much *contributors* as they are *mediators* in the project of representing the physical world around us. But more importantly, they neither merely execute fundamental theory nor simply accommodate empirical phenomena. Rather, as the example of many-body models in condensed matter physics demonstrates, they are highly structured entities which are woven into, and give stability to, scientific practice.

---

[14] A model whose predictions of the order parameter are *systematically* wrong (e.g., consistently too low) but which gets the qualitative behaviour right (e.g., the structure of the phase diagram), may be preferable to a model that is more accurate for most situations, but is vastly (qualitatively) mistaken for a small number of cases. Likewise, a model that displays certain symmetry requirements or obeys certain other rigorous relations may be preferable to a more accurate model (with respect to the *physical* observables in question) that lacks these properties.

[15] See also Sect. 11.5.2; for a full case study of cross-model transfer of warrant, see (Gelfert 2009).

References

Bailer-Jones, D.M.: When scientific models represent. Int. Stud. Philos. Sci. **17**, 59–74 (2003)

Batterman, R.: Asymptotics and the role of minimal models. Br. J. Philos. Sci. **53**, 21–38 (2002)

Baxter, R.J.: Exactly Solved Models in Statistical Mechanics. Academic Press, New York (1982)

Bohr, N.: Studier over metallernes elektrontheori, Københavns Universitet. Repr.: The Doctor's Dissertation (Text and Translation). In: Rosenfeld, L., Nielsen, J. Rud. Early Works (1905-1911). Niels Bohr Collected Works 1 (1911)

Boumans, M.: Built-in justification. In: Morrison, M., Morgan, M.S (eds.) Models as Mediators. Perspectives on Natural and Social Science, pp. 68–96. The MIT Press, Cambridge (1999)

Brush, S.: History of the Lenz-Ising model. Rev. Mod. Phys. **39**, 883–893 (1967)

Cartwright, N.: The Dappled World. A Study of the Boundaries of Science. Cambridge University Press, Cambridge (1999)

Cartwright, N: Models and the limits of theory: quantum hamiltonians and the BCS model of superconductivity. In: Morrison, M., Morgan, M.S (eds.) Models as Mediators: Perspectives on Natural and Social Science, pp. 241–281. Cambridge University Press, Cambridge (1999)

Fischer, M.E.: Scaling, universality, and renormalization group theory. In: Hahne, F.J.W. (ed.) Critical Phenomena (Lecture Notes in Physics, vol. 186), pp. 1–139. Springer, Berlin (1983)

Galam, S.: Rational group decision-making: a random-field Ising model at $T = 0$. Phys. A **238**, 66–80 (1997)

Gebhard, F: The Mott Metal–Insulator Transition: Models and Methods (Springer tracts in modern physics, vol. 137), Springer, Berlin (1997)

Gelfert, A.: Rigorous results, cross-model justification, and the transfer of empirical warrant: the case of many-body models in physics. Synthese **169**, 497–519 (2009)

Gelfert, A.: Mathematical formalisms in scientific practice: From denotation to model-based representation. Stud. Hist. Philos. Sci. **42**, 272–286 (2011)

Gelfert, A.: Strategies of model-building in condensed matter physics: trade-offs as a demarcation criterion between physics and biology? Synthese **190**, 253–272 (2013)

Gibson, M.C: Implementation and application of advanced density functionals. Ph.D dissertation, University of Durham (2006)

Giere, R.N.: Using models to represent reality. In: Magnani L, Nersessian, N.J., Thagard, P. (eds.) Model-Based Reasoning in Scientific Discovery, pp. 41–57. Plenum Publishers, New York (1999)

Goldenfeld, N.: Lectures on Phase Transitions and the Renormalization Group (Frontiers in Physics, vol. 85). Addison Wesley, Reading (1992)

Griffiths, R.B.: Rigorous results and theorems. In: Domb, C., Green, M.S. (eds.) Phase Transitions and Critical Phenomena, pp. 8–109. Academic Press, New York (1972)

Heisenberg, W.: Theorie des Ferromagnetismus. Zeitschrift für Physik **49**, 619–636 (1928)

Hughes, R.I.G.: The Ising model, computer simulation, and universal physics. In: Morrison, M., Morgan, M.S (eds.) Models as mediators. Perspectives on Natural and Social Science, pp. 97–145. Cambridge University Press, Cambridge (1999)

Ising, E.: Beitrag zur Theorie des Ferromagnetismus. Zeitschrift für Physik **31**, 253–258 (1925)

Knuuttila, T.: Some consequences of the pragmatist approach to representation: decoupling the model-target dyad and indirect reasoning. In: Suárez, M., Dorato, M., Rédei, M. (eds.) EPSA Epistemology and Methodology of Science, pp. 139–148. Springer, Dordrecht (2010)

Krieger, M.H.: Phenomenological and many-body models in natural science and social research. Fundamenta Scientiae **2**, 425–431 (1981)

Matsuda, H.: The Ising model for population biology. Prog. Theoret. Phys. **66**, 1078–1080 (1981)

Morrison, M.C.: Modelling nature: between physics and the physical world. Philosophia Naturalis **35**, 65–85 (1998)

Morrison, M.: Models as autonomous agents. In: Morrison, M., Morgan, M.S. (eds.) Models as Mediators. Perspectives on Natural and Social Science, pp. 38–65. Cambridge University Press, Cambridge (1999)

Morrison, M., Morgan, M.S. (eds.) Models as Mediators. Perspectives on Natural and Social Science. Cambridge University Press, Cambridge (1999a)

Morrison, M., Morgan, M.S. (eds.): Models as mediating instruments. In: Morrison, M., Morgan, M.S. (eds.) Models as Mediators. Perspectives on Natural and Social Science, pp. 10–37. Cambridge University Press, Cambridge (1999b)

Niss, M.: History of the Lenz-Ising model 1920–1950: from ferromagnetic to cooperative phenomena. Arch. Hist. Exact Sci. **59**, 267–318 (2005)

Nozières, P.: Theory of Interacting Fermi Systems. Benjamin, New York (1963)

Ogielski, A.T., Morgenstern, I.: Critical behavior of 3-dimensional Ising model of spin glass. J. Appl. Phys. **57**, 3382–3385 (1985)

Onsager, L.: Crystal statistics. I. A two-dimensional model with an order–disorder transition. Phys. Rev **65**, 117 (1944)

Suárez, M.: Theories, models, and representations. In: Magnani, L., Nersessian, N.J., Thagard P. (eds.) Model-Based Reasoning in Scientific Discovery, pp. 75–83. Plenum Publishers, New York (1999)

Tusch, M.A., Szczech, Y.H., Logan, D.E.: Magnetism in the Hubbard model: an effective spin Hamiltonian approach. Phys. Rev. B **53**(9), 5505–5517 (1996)

van Leeuwen, H.J.: Problèmes de la théorie électronique du magnétisme. Journal de Physique et le Radium **2**(12), 361–377 (1921)

Yi, S.W.: The nature of model-based understanding in condensed matter physics. Mind Soci. **5**, 81–91 (2002)

Chapter 12
How Do Quasi-Particles Exist?

Brigitte Falkenburg

Quasi-particles emerge in solids. They are excitations of a macroscopic many-particle system. Such an excitation is a quantum effect, consisting in the addition of a quantum of energy to the quantum state of the solid. Hence, quasi-particles are collective quantum effects of all the charges or nuclei of a macroscopic atomic lattice such as a crystal. Under certain conditions they are separable and localizable. In this case, they behave like free (i.e., unbound, uncoupled) subatomic particles, and for this reason they are called quasi-particles. But how do they exist? What is their ontological status? Are they as real as electrons or protons, or not? In the context of the debate on scientific realism, the concept of quasi-particles is puzzling. Given that they do not exist on their own but only as collective effects, they seem to be fake entities rather than physical particles. But they can be used as markers, etc., in crystals. Hence, it is possible to use them as technological tools, even though, taken on their own, they are not entities. It has been argued that, for this reason, they counter Hacking's reality criterion, "If you can spray them, they exist." But as I will show, this line of reasoning misses the crucial point that quasi-particles are real collective effects which contribute to the constitution of a solid.

In order to spell out the ontological status of quasi-particles, I proceed as follows. (1) I give a rough sketch of the issue of scientific realism and (2) classify the particle concepts as they stand in physics today. In the context of current quantum theories, *several* particle concepts coexist and *none* of them reduces to a classical particle concept. Quantum particles have some but not all the hallmarks of classical particles. In addition, non-local wave-like properties must be attributed to them. In a certain sense, this matter of fact is neglected in the debate on scientific realism. (3) I then consider quasi-particles, in particular, the underlying theory and their particle properties. It is instructive to compare them with the field quanta of a quantum field, on the one hand, and subatomic matter constituents, on the other. The comparison with virtual particles (Sect. 12.3.4) and quarks (Sect. 12.3.5) is of particular interest, given that neither of these kinds of field quanta can be individuated (for quite different reasons).

B. Falkenburg (✉)
Faculty of Human Sciences and Theology, Department of Philosophy and Political Science,
TU Dortmund, 44221 Dortmund, Germany
e-mail: brigitte.falkenburg@tu-dortmund.de

© Springer-Verlag Berlin Heidelberg 2015
B. Falkenburg and M. Morrison (eds.), *Why More Is Different*,
The Frontiers Collection, DOI 10.1007/978-3-662-43911-1_12

All these quantum particle concepts differ substantially from the classical particle concept. (4) Finally, I discuss the question of whether quasi-particles do in fact go against Hacking's reality criterion, as Gelfert has argued. (5) I conclude that quasi-particles are genuine quantum entities, which are as real or unreal as electrons, protons, quarks, or photons, even though all these quantum entities have quite different characteristics.

12.1 Scientific Realism

The current debate on scientific realism deals with old philosophical problems that come in new clothes. The debate is very old, it dates back to the beginnings of modern science. In Galileo's day, Aristotelians claimed that astronomy aims merely at saving phenomena. They defended instrumentalism in the debate about competing world systems. Galileo, however, claimed that science aims at truth, and he defended the truth of the Copernican system. The success of classical physics is due to a belief in the existence of laws of nature and unobservable entities such as atoms, forces, and fields. But when the atoms became the subject of physical theory, instrumentalism came back. At the end of the 19th century, Mach claimed that physical theories just aim at economy of thought, while defending empiricism and attacking atomism. One of his reasons was that the classical models attributed weird properties to the atoms.[1] Even atomists like Maxwell and Boltzmann kept a certain instrumentalistic attitude towards their kinetic theory, doubting whether the theory gave a true account of physical reality.[2] When Planck started his work on thermodynamics and radiation theory, he was a follower of Mach. But in order to reconcile electrodynamics with the second law of thermodynamics, he made a shift toward Boltzmann's statistical account of entropy. He introduced his quantum of action h into the theory of black body radiation and converted to atomism. In 1908, he defended a very strong version of scientific realism against Mach.[3]

However, the rise of quantum theory showed that in one crucial respect Mach's distrust of classical atoms had been on the right track. According to quantum mechanics, there are no *classical* atoms, even though there are *atoms* which consist of subatomic particles. In view of the missing electron orbits and the quantum mechanical many-particle wave function of the electrons inside the atom, the realism debate changed topics. With the Bohr–Einstein debate, it shifted to the interpretation of quantum mechanics. Even though the old opposition of instrumentalism and realism could not really cope with the new situation, up to the present day it is usual to interpret the opposing positions of Bohr and Einstein in

---

[1] Mach 1883, p. 466.

[2] See Maxwell 1859 and the Boltzmann-Zermelo debate (Ehrenfest 1911); see also Scheibe 2001, pp. 145–148, and Scheibe 2007, pp. 100–103.

[3] Planck 1908; see also Scheibe 2001, pp. 148–151, and Scheibe 2007, pp. 55–69.

such terms. Bohr, however, emphasized the non-separability of the quantum system and the measuring device as well as the holistic features of quantum phenomena, whereas Einstein insisted on the independence and separability of subatomic particles at a space-like separation. Their views about the existence of quantum systems did not differ. The disagreement was rather about the context-dependence or -independence of their states, or about the non-locality or locality of physical reality, respectively. Since the 1980s, quantum entanglement and the violation of Bell's inequality are well-established phenomena, which are in particular investigated in the experiments of quantum optics. Today, local realism stands for Bell's inequality and its violation stands for non-locality. Whether some version of non-local realism is tenable is still under debate. Bohr's version of non-local realism has never been spelled out, or only along the lines of Einstein's realism, in the form of Bohm's non-local hidden variable theory.[4] The ontological status of subatomic particles has remained unclear up to the present day. This matter of fact should be kept in mind in the following discussion of the ontological status of quasi-particles, which does not differ significantly from the ontological status of subatomic particles in general.

The current debate on scientific realism has not contributed to clarifying these issues. It was opened by Kuhn's claim that, due to scientific revolutions, there is no stable scientific truth, and hence no true account of physical reality.[5] In the post-Kuhn era, the debate focused on the existence of unobservable entities and the truth of the laws of physics. The philosopher Maxwell argued that the observational tools of physics such as telescopes, the magnifying glass, the microscope and electron microscope, and particle accelerators allow observation of objects of all sizes which cannot be seen with the naked eye, from distant galaxies to subatomic particles.[6] Against Maxwell, van Fraassen defended a neo-Machian version of empiricism, according to which we have no knowledge of unobservable entities.[7] Nancy Cartwright raised objections against the truth of the laws of fundamental theories, and supported entity realism about causal powers or capacities of nature.[8] At the same time, Hacking brought up his famous 'technological' reality criterion: "If you can spray them, they exist".[9] More positions were developed, up to Worrall's structural realism, the current version of realism about the laws of physics.[10] From 1962 up to the present day, the debate has proceeded as if there were no quantum non-locality. Maxwell's argument in favor of the observability of distant galaxies and microscopic entities assumes that these entities are local, hence it supports some version of local realism. Indeed this point is crucial for the argument about quasi-particles and Hacking's reality criterion discussed below.

---

[4] See Bohm 1952.

[5] Kuhn 1962.

[6] Maxwell 1962.

[7] Van Fraassen 1980.

[8] Cartwright 1983, 1989.

[9] Hacking 1983, pp. 22–25.

[10] Psillos 1999, Worrall 1989.

12.2 Particle Concepts

Before discussing the quasi-particles, let me give a rough sketch of the particle concept(s) of current quantum physics and their relation to the issues of scientific realism. Current physics embraces a wide variety of particle concepts. There are as many particle concepts as there are physical theories, or even more. Contrary to Kuhn's claims about the results of scientific revolutions, the theories of classical physics did not die off in favor of a universal quantum world view. They survived, given that up to the present day there is no unified theory of the classical and the quantum domain, and hence no unified world view of physics. Current physics is a patchwork of laws from classical and quantum theories, correspondence rules, and meta-theoretical bridge principles such as symmetries. Some theories of physics deal with classical particles, others with various kinds of quantum particles. In all cases, the particle concepts are theory- (or model-) dependent. In a first approach, the variety of particle concepts may be roughly classified as follows:[11]

(CP)   *Classical concept*: Particles have local states. Their properties are completely determined by their dynamics. They propagate along non-intersecting spatio-temporal trajectories, due to which they are individuated (classical statistics). Their dynamic properties are mass, momentum/energy, and charge.

(QP)   *Quantum concept*: Particles have non-local states, but may be approximately localized by means of a position measurement. Their properties are determined by a deterministic wave dynamics, from which probabilistic predictions for individual measurement outcomes derive in terms of quantum mechanical expectation values. They propagate like waves, and they are indistinguishable (Fermi/Bose statistics). Their dynamic properties are mass, momentum/energy, and charge; in addition, spin and parity.

(OP)   *Operational concept*: Particles are the local events in a measuring device for position measurements, such as a Geiger counter, the Wilson chamber, or a modern particle detector. Their properties are what is measured by the particle detector, i.e., mass, momentum/energy, and charge; in addition, spin and parity (depending on the measuring device).

These concepts are related as follows. Strict versions of (CP) and (QP) are incompatible. Only under certain conditions (e.g., in quantum mechanics in phase space) can (QP) give rise to an approximate, probabilistic version of (CP). On the other hand, (OP) is compatible with both (CP) and (QP). (OP) only refers to what is measured by means of a particle detector. (OP) is, so to speak, the empiricist version of both (CP) and (QP). Scientific realists (no matter what specific version of realism) and empiricists, or instrumentalists, cannot disagree about (OP), but only

---

[11] For a more precise overview, see Falkenburg 2012. For all details, see Falkenburg 2007, Chap. 6.

about the non-empirical features of (CP) and/or (QP), namely, the particle trajectory and/or wave propagation between the measurement points of a particle track. From the viewpoint of physics, (CP) and (QP) still have in common that the dynamic properties of the respective particle concepts underlie certain conservation laws. The conservation laws for mass/energy, spin, parity, and various kinds of charges give rise to corresponding symmetries, which in turn allow one to identify particles with the irreducible representations of symmetry groups.[12] Hence today, the particle concept is usually defined in terms of symmetry groups:

(SG)   *Symmetry concept*: (Elementary) particles are the (irreducible) representations of symmetry groups. Their dynamic properties are the parameters according to which the representations are classified: mass (or energy, respectively), spin, and parity (Lorentz group of Special Relativity); and the generalized charges of flavor (U(1) × SU(2) of the electroweak interaction) and color (SU(3) of the strong interaction).

(SG) is compatible with the three former particle concepts. The parameters according to which the irreducible representations of symmetries are classified correspond to the measured properties of (OP), on the one hand, and the theoretical properties (QP) or (CP), on the other, depending on whether the underlying dynamics is quantized or not or not.[13] So far, and according to a modest ontological approach, particles seem to be bundles of dynamic properties (mass/energy, spin, parity, and various kinds of charges) which underlie conservation laws, and they may be grouped according to the corresponding symmetries.

However, life is not so easy. (SG) only deals with non-interacting, unbound, uncoupled, or 'free', particles, whereas (CP) and (QP) also describe interacting particles. According to (CP) and (QP), particles may form compound systems. And according to the respective underlying dynamics, (OP) is due to the interactions of a classical particle or a quantum system with a measuring device.

At this point, more kinds of quantum particles come into play: the field quanta of a quantum field, embracing real and virtual particles, matter constituents, and finally, quasi-particles. In order to compare them, let me sketch the corresponding concepts. All of them are subspecies of (QP).

(FQ)   *Field quanta*: The field quanta of a quantum field theory are in non-local states, propagate like waves, and are subject to a probabilistic dynamics which is expressed in terms of creation and annihilation operators. They are indistinguishable (Fermi/Bose statistics). Their dynamic properties are mass ≥ 0, momentum/energy, and the generalized charges of flavor and color; in addition, spin and parity.

---

[12]   Wigner 1939.

[13]   It should be noted, however, that Wigner 1939 does not deal with particles, but with the solution of field equations.

(FQ$_R$)   *Real (physical) particles* These correspond to the incoming and outgoing
           quantum waves of scattering processes. They may be approximately
           localized by means of a position measurement. The detection of a particle
           corresponds to the annihilation of a field quantum.

(FQ$_V$)   *Virtual particles*: These correspond to the Feynman diagrams which
           contribute to the perturbation expansion of the S-matrix of the interacting
           quantum fields. Their dynamic properties may be 'off-shell', i.e., they
           may violate energy conservation during the interaction.

(QP$_M$)   *Matter constituents*: These may form compound systems. The parts of the
           whole are in entangled quantum states and indistinguishable (Fermi/Bose
           statistics). Their dynamic properties underlie sum rules that derive from
           the corresponding conservation laws.

(QP$_Q$)   *Quasi-particles*: These are the excitations of a solid. Their dynamic
           properties are an effective mass, momentum/energy, charge, and/or spin.
           These properties underlie sum rules that derive from the corresponding
           conservation laws.

All these quantum particles have non-local features. They propagate like waves
and may be at most approximately localized by means of a position measurement.
Hence from the remarks of the last section it should be clear that local realism
cannot cope with them. What else can be said about them, with regard to scientific
realism? Obviously, the associated concepts differ in operational content. Only the
real field quanta of a quantum field, or their annihilation, correspond to the operational concept (OP), that is, the detection of particles. Therefore, the real field
quanta are called 'physical particles'. The rest is theory, namely, the mathematical
wave function which is a solution of the corresponding quantized field equation(s).
The physical particles as such are observed in position measurements. Their
propagation before and/or after their detection is inferred from the corresponding
conservation laws.

In the practice of physics, however, the operational content of (FQ$_R$), or the
concept of physical particles, is usually understood in a broader sense. The experiments of quantum optics and particle physics are performed with beams of electrons,
photons, etc., of a given momentum or energy, respectively. Generated by a laser or a
particle accelerator, the beams are prepared in a well-defined quantum state of sharp
momentum or energy. In the experiments of quantum optics with 'single' particles,
the beam is in addition prepared in a quantum field state with occupation number 1.
For physicists, the preparation of a quantum state by means of an experimental device
has the same operational meaning as the detection of a particle. This matter of fact is
important since in most cases the preparation gives rise to plane quantum *waves* (i.e.,
quantum systems in a non-local state of well-defined momentum and energy),
whereas detection gives rise to quantum *particles* (i.e., quantum systems in an

approximately local state of well-defined position).[14] In particle physics, the physical particles corresponding to (FQ$_R$) are the incoming and outgoing beams of a scattering experiment. In quantum optics, the physical particles or real field quanta are the quantum states prepared and detected by means of the experimental arrangement. For physicists, the quantum waves prepared by a laser, an electron gun, a particle accelerator, etc., and the quantum particles detected by a Geiger counter, Wilson chamber, photomultiplier, etc., are completely on a par with regard to their ontological status. Hence they consider (FQ$_R$) to have broader operational content than (OP). For them, there exists an operational wave concept (OW), too:

(OW)   *Operational wave concept:* Quantum waves (or their squared amplitude) are the beams prepared in a field state of well-defined momentum/energy and occupation number by means of an appropriate experimental device, such as a laser, an electron gun, a particle accelerator, etc. Their properties are the momentum/energy, intensity, polarization, etc., measured in a calibration measurement.

For physicists working in particle physics, quantum optics, or condensed matter physics, quantum waves and quantum particles are ontologically on a par. The Nobel prize winner Wolfgang Ketterle once stressed this quantum pragmatism in a popular talk about the Bose–Einstein condensate. He told the public that it is hard to understand quantum mechanics, but after several years of physical practice one gets used to *preparing waves* and *detecting particles*.[15] In this nice statement, the experimental procedures of preparation and measurement are both understood as empirical procedures, but with different outcomes. What makes the difference is the quantum state. The preparation of a beam aims at a plane wave, the detection of a particle produces an approximately local state.[16] The beam prepared in a quantum experiment obviously satisfies Hacking's criterion "When you can spray them, they are real", given that it is prepared in order to perform a measurement. Indeed the preparation of a quantum beam is made precisely in order to guarantee manipulative success.

The debate on scientific realism, however, cannot cope with such quantum pragmatism, according to which non-local and approximately local quantum states are ontologically equal. A full-blooded empiricist would say that the existence of quantum particles or waves is only granted when they are detected. This view supports (OP) and an ensemble interpretation of the quantum mechanical wave function, but not it does not support (OW). A local realist would ask for hidden variables in order to obtain more of physical reality. But in the context of a relativistic quantum field theory, which in particular predicts large fluctuations in the occupation number of a quantum field, a hidden-variable interpretation of quantum

---

[14] For a discussion of the distinction between preparation and detection, and wave–particle duality in physical practice in general, see Falkenburg 2007, Chap. 7.

[15] Talk given at the annual meeting of the German Physical Society in Hannover, March 2003.

[16] For more details, see Falkenburg 2007, Sect. 7.3.

theory is hard to come by. Quantum pragmatism follows Bohr into a middle position between empiricism and realism. According to this, the conservation laws of physics grant that the prepared quantum waves exist and propagate through the apparatus, and the non-local correlations of their detections demonstrate the holistic features of quantum phenomena.

Let us now look at the other quantum particle concepts in the context of scientific realism, starting with *virtual particles*. All sides of the debate should agree that the virtual particle concept has no direct operational content. Virtual particles are called 'virtual' since they are not subject to (OP). Empiricists and followers of Bohr would doubt that *single* virtual particles actually exist. Given that the propagators of virtual particles give rise to the partial amplitudes of a perturbation series, their ontological status seems to be similar to Ptolemy's epicycles. It reduces to the instrumental character of mathematical auxiliary functions which stem from the observer's point of view, but do not correspond to any real structures of the world. (The propagators of virtual particles are needed by the classical observer in order to describe the interactions of quantum fields, just as epicycles and deferents are needed by the terrestrial observer in order to describe the solar system). Local realists would probably disagree, insisting on the existence of single virtual particles. Finally, empiricists and the followers of Bohr would perhaps disagree about the question of whether the sum total of all virtual particles gives rise to collective effects such as the Lamb shift, the Casimir effect, etc. Followers of Bohr would insist on the holistic features of physical reality which cause such collective effects, whereas empiricists would prefer causal agnosticism.

Second, what about *matter constituents*? Physics tells us that quarks, electrons, protons and neutrons are dynamic parts of the atoms of which macroscopic bodies consist. The constitution of bodies underlies the conservation laws for mass, energy, and so on, which are well-established by the experiments of atomic, nuclear, and particle physics. In the debate on scientific realism, the question of whether and to what extent the constituent models of matter represent physical reality has not yet been discussed in detail. In my "Particle Metaphysics", I defend the view that, according to the sum rules for dynamic properties, the parts of a given whole exist, even if they are in entangled quantum states.[17] The electrons, protons, neutrons, and quarks inside the atom belong to an entangled bound quantum state and so cannot be spatio-temporally individuated. Nevertheless, the conservation laws of physics hold for them, giving rise to sum rules for mass/energy, momentum, spin, and charges, according to which the matter constituents are *dynamic* parts of matter. The corresponding conservation laws and sum rules underlie the way in which the protons and neutrons are made up of quarks according to the quark model, just as they underlie the laws of nuclear fission and fusion. An anti-realist about quarks may claim that only particles which can be detected on their own outside the atom, such as electrons, protons, neutrons, muons, or pions, genuinely exist, whereas no genuine particles

---

[17] See Falkenburg 2007, Sect. 6.5.

exist inside an entangled quantum system; whence, due to quark confinement, quarks do not genuinely exist. Any particle physicist would counter that the scattering experiments of high energy physics measure the structure functions of the proton and neutron, and with them the way in which the mass, charge, spin, and magnetic momentum of the proton and neutron add up from corresponding dynamic properties of the quarks. Hence, the case for realism about quarks is something like Putnam's miracle argument: Where should the specific mass, charge, spin, and magnetic momentum of the proton and neutron come from, if not from corresponding dynamic properties of genuinely existing entities such as quarks?

And finally, *quasi-particles*? Let us see! Two points may already be noted. The concept is decisively based on the sum rules just mentioned. And in order to clarify their ontological status, we will need to compare them with virtual particles and matter constituents, and their respective ontological status.

12.3 Quasi-Particles

Quasi-particles are charged or uncharged energy quanta in a solid. They propagate through a solid and interact with each other *as if* they were single particles, in a sense that needs to be qualified. They result from a collective excitation of *many* subatomic constituents of a solid. Here, 'many' means a collection of the order of 6×10^{23} electrons and atomic nuclei.[18] Let me start with the theory, then compare the concept of quasi-particles with the other particle concepts sketched above, and finally discuss their ontological status and its relation to Hacking's reality criterion.[19]

12.3.1 The Theory

The basic theory is the quantum mechanics of a many-particle system. The constituents of a solid are electrons and atomic nuclei. The electrons and atomic nuclei of the solid are matter constituents in the sense of the above particle concept (QP_M). The solids are classified into metals and various kinds of crystals. The latter are distinguished in terms of chemical binding on the one hand, and conductors, semiconductors, or insulators on the other.

Many aspects of the behavior of conductors, semiconductors, and insulators can be described by means of well-established quantum mechanical approximation methods. The simplest approximation procedure is the Hartree–Fock theory, also called the self-consistent field or mean-field approximation. It aims to describe the

---

[18] For the following, see Anderson 1997.

[19] The following sections are based on Falkenburg 2007, Sect. 6.4.4, which gives a much more detailed account.

behavior of one electron within the field generated by all the other charges of the constituents of a solid. This gives rise to the well-known band structure of solids (which underlies the model of a transistor, for example). The theory (or model) assumes that a single electron moves independently in the field of all the other electrons and nuclei of the solid. This approximation is based on the collective effects that give rise to the appearance of the kinds of excitations called quasi-particles. In general, such effects are so large that they cannot be treated as perturbations. Hence, the Hartree–Fock approximation has the typical features of an idealization which gives an approximately true description of physical reality in some regards, while going crudely wrong in others. Usually the Hartree–Fock Hamiltonian does not yield a correct description of the ground state of a solid, i.e., the quantum state of lowest energy. The best that can be achieved in this way is a surprisingly good approximation to the ground state energy.

However, the most important quantity associated with a solid is not the ground state energy but the energy difference between the ground state and the lower excited states.[20] It determines many macroscopic phenomenological features of a solid, and in particular its magnetic properties. In order to calculate this crucial energy difference, the methods of quantum field theory are used. There is a far-reaching formal analogy between quantum field theory and the many-particle theory of solid state physics. The analogy is based on the fact that the so-called second quantization (the formal quantization of the many-particle Schrödinger wave function) and the quantization of a classical field give the same formal results. Since the starting point is a non-relativistic field equation, the result is a non-relativistic analogue of quantum field theory.

Even in the relatively simple Hartree–Fock approach the analogy is very helpful. It permits one to get rid of the complicated antisymmetrized many-electron wave functions of a solid and replace them by the number representation of the field modes of a quantum field in Fock space.[21] Electrons are indistinguishable and obey Pauli's exclusion principle, in accordance wirh (QP). Hence, all that matters is the occupation number 0 or 1 of the quantum states of the electrons. The formalism is based on the creation and annihilation operators a_k^+, a_k and the commutation rules for field quanta:

$$[a_k, a_k^+] = 1$$

$$[a_k, a_k] = 0$$

$$[a_k a_{k'} \pm a_{k'} a_k] = 0$$

The anti-commutator $[a_k a_{k'} + a_{k'} a_k]$ applies to fermions, the commutator $[a_k a_{k'} - a_{k'} a_k]$ to bosons. The operators act on the field modes. The creation

---

[20] Anderson 1997, pp. 97–99.
[21] See Anderson 1997, pp. 15–28.

operator a_k^+ generates a field quantum of wave number k, while the annihilation operator a_k makes a field quantum of the corresponding field mode disappear:

$$a_k^+ \Psi_{k,n} = \sqrt{n+1} \; \Psi_{k,n+1}$$

$$a_k \Psi_{k,n} = \sqrt{n} \; \Psi_{k,n-1}$$

The field state Ψ is expanded into a sum of field modes $\Psi_{k,n}$ with wave number k and well-defined occupation numbers n (these states form the Fock space):

$$\Psi = \Sigma_{k,n} \, c_{k,n} \Psi_{k,n}$$

The states in this sum are eigenstates of the occupation number operator $N = a_k^+ a_k$:

$$N \Psi_{k,n} = n \Psi_{k,n}$$

If the creation operator a_k^+ acts on the vacuum state Ψ_0, a field mode $\Psi_{k,1_k}$ with one field quantum of wave number k is generated. The corresponding eigenstate of the occupation number operator is

$$N \Psi_{k,1} = 1 \; \Psi_{k,1}$$

Electrons are fermions. Any electron is represented by a state $c_k c_k^+ \Psi_k$ that belongs to the occupation number 1 in the Fock space. According to this approach, approximately uncoupled electrons have the properties of a Fermi gas.

Quasi-particles enter where the Hartree–Fock approximation becomes too crude. As mentioned above, although the Hartree–Fock approach gives a surprisingly good approximation of the ground state *energy* of a solid, the most important quantity is not the ground state energy but the energy *difference* between the ground state and the lower excited states. These states are due to *collective excitations* of the electrons and/or nuclei of the solid. Collective excitations concern the solid as a whole, a collection of the order of 6×10^{23} electrons and atomic nuclei. This means that they are *macroscopic quantum states* of a solid. In order to model them, the Hartree–Fock approach is no longer useful and the methods of quantum field theory, in particular, the occupation number formalism, are employed.[22] Most

---

[22] This approach is strikingly similar to the application of the formalism of quantum field theory to bacteria which is explained by Meyer-Ortmanns in her contribution to this book (Sect. 2.2.4). In both cases the approach is phenomenological, making use of a formal analogy between the phenomena under investigation and the phenomena of particle creation and annihilation described by quantum field theory. In the case of quasi-particles, however, the analogy extends to the dynamic properties of the entities which are created and annihilated.

important are the lower excited states mentioned before. They are called *elementary excitations*. In a first approach, they can be defined as field quanta, in terms of creation and annihilation operators q_k^+, q_k for quanta of a given momentum k:

> This, then, is a preliminary definition we can make of an elementary excitation: *An elementary excitation of momentum k is that operator which creates the lowest excited state of a particular type of momentum k from the ground state.*[23]

It should be noted that Anderson calls the definition "preliminary" because the quasi-particle concept is based on an *approximation procedure*. In a second step, the interactions *between* quasi-particles have to be taken into account (see below). The excited states are "of a particular type" regarding the various *kinds* of excitation. These may be due to adding an electron, the slow vibrations of the ionic atom cores of a solid, spin fluctuations, optical excitations, etc. The lowest excited energy states which result from these different elementary excitations may differ substantially. They are calculated in perturbation theory. The approximation procedure is analogous to the perturbation expansion of quantum field theory, including the use of Feynman diagrams. To zero order, each quasi-particle is considered independently. To first order, their interactions are calculated in the Born approximation.

12.3.2 The Concept

The calculation shows that the elementary excitations have *approximate* particle properties in the sense of (QP). In particular, they have an approximate charge and/or spin, they are fermions or bosons, they behave independently, i.e., approximately like non-interacting or uncoupled particles, and they have some effective mass m_e. All these properties are approximate in the sense of perturbation theory, as explained above. To zero order, quasi-particles are considered to be independent. To first order, their interactions are calculated in the Born approximation (see Sect. 12.3.3). Since this feature of their description may give rise to the objection that quasi-particles have no genuine, 'really' independent existence, let me note this: in a certain sense the same is true of planets, or the charged particles in a scattering experiment. To zero order, planets are inertial masses that move independently of the rest of the universe on an inertial trajectory. To first order, they interact with the sun according to Kepler's laws, which approximately derive from Newton's law of gravitation; and to higher orders, with the other bodies of the solar system in a full-fledged perturbation expansion of celestial mechanics. In a scattering experiment, the scattered 'probe' particles are considered to be independent before the scattering, just as the particles

---

[23] Anderson 1997, p. 102. Anderson's momentum k is the quantity called $p = \hbar k$ in Falkenburg 2007.

which are detected after the scattering. According to the quantum mechanics of scattering, they are modeled as *approximately* free long before and long after the interaction. The same holds for their classical analogue, the α-particles of Rutherford's famous scattering experiment.[24] Hence, if quasi-particles have no genuine independence according to the approximations employed in the model, then planets, and indeed the α-particles of Rutherford's scattering experiment, likewise have no genuine independence.

From this point of view, the main difference between planets and electrons (or classical and quantum particles), on the one hand, and quasi-particles, on the other, is not their approximate independence of each other, but the way in which the latter emerge in solids. Due to collective excitation, they emerge in solids as quasi-independent entities in their own right with certain effective dynamic properties. In contradistinction to the mass of planets or free electrons, their dynamic properties depend on their environment, the solid. Without the solid, there are no quasi-particles. In *this* (and *only* this) sense, they do not come on their own. Because they behave approximately independently *within* the solid and due to their effective dynamic properties, they are called *quasi-particles*. The resulting concept (QP$_Q$) has already been given:

(QP$_Q$)   *Quasi-Particles* are the excitations of a solid. Their dynamic properties are an effective momentum/energy, mass, charge, and/or spin, which underlie sum rules deriving from the corresponding conservation laws

Quasi-particles have non-local states, but they may be approximately localized under certain conditions. Their properties are determined by many-particle quantum mechanics in Hilbert space or non-relativistic quantum field theory in Fock space. Quasi-particles propagate like waves, and they are indistinguishable (Fermi/Bose statistics). Their effective dynamic properties derive from the sum rules for the solid as a whole, according to the conservation laws of physics.

The *effective mass* m_{eff} of a quasi-particle is interaction-dependent. It is associated with its energy E, which is given by the energy difference between the excited state and the ground state of the solid. The effective mass is due to a collective effect that is analogous to mass renormalization in quantum field theory. As in quantum field theory, physicists like to say that quasi-particles are 'dressed'. In solid state physics, this ontological *façon de parler* seems to be better justified, since the effective mass is due to the interactions within a solid. (However, this point is debatable. The analogy between the field quanta of a quantum field and the quasi-particles of solid state physics induces a formal analogy between the vacuum state of a quantum field and the ground state of a solid. And the Casimir effect shows that the vacuum state of a quantum field gives rise to observable phenomena.)

---

[24] For Rutherford scattering, there is indeed an exact formal correspondence between the classical and the quantum case. This matter of fact underlies the definition of form factors in nuclear and particle physics; see Falkenburg 2007, pp. 136–137.

There are many different kinds of quasi-particles. Some of them are localizable, others are not. Particle-like properties in the sense of (QP$_Q$) are attributed to all of them. In particular, the following three kinds of quasi-particles are well-known:[25]

1. *Free electronic charge carriers.* These result from adding an electron to a solid. They inherit their charge and spin from this electron, but their effective mass depends on their velocity. In a metal, the additional electron "disappears" in the partially filled band of conduction valence electrons, i.e., its momentum distributes within the solid, being carried away by the band electrons and giving rise to a vanishing excitation energy. In a semiconductor or insulator, the wave function of the added electron develops in a different way. After a certain time, a wave packet remains. The wave packet is centered around a central momentum value k_0 (which corresponds to the excitation energy) and has a width Δk. It is then localized within a region of size $1/\Delta k$ around some point r_0. This wave packet approximately carries the charge and spin of the electron. Hence, it behaves like a fermion.

2. *Phonons.* Like other density waves such as magnons or plasmons, these result from collective excitations of the nuclei. Phonons result from the slow vibrations of the ionic cores mentioned above. They are stationary density waves and as such they do not inherit any dynamic properties from physical particles. In particular, they carry no charge and no spin, being approximately bosons. Phonons behave like the harmonics of sound waves (hence their name). In many solids they are not localized. However, in crystals they may even be approximately localized at impurities and imperfections.

3. *Excitons.* Excitons occur in insulators. They are due to optical excitation. An electron of an ionic atom core in a crystal such as GaAs (which is of great importance in telecommunications) is excited from a predominantly arsenide p state to a predominantly gallium s state. The excited electron can move in the p-band through the crystal. Due to Coulomb interaction, it repels the s-electrons of the atoms it passes. Therefore, it draws its corresponding hole along behind it in the s-band. Such an electron–hole pair is a bound state of an electron and a hole that gives rise to a compound quasi-particle of charge 0 with a certain effective mass and with bosonic behavior. Like free electric charges and phonons, they may be localized under certain circumstances. Spin waves (which give rise to the magnetic state of a solid) and the Cooper pairs of superconductivity are similar to excitons.

12.3.3 Comparison with Physical Particles

To sum up, quasi-particles are excitations of a solid which *behave approximately* like particles, even though they *are not* particles in the sense of local or localizable entities existing outside a solid. The "quasi" in their name is due to the matter of

---

[25] See Anderson 1997, pp. 102–104.

fact that they do not come on their own, but stem from collective excitations of a solid. They look *as if* they come on their own, but they do not. They are context-dependent. Their effective dynamic properties, such as charge and mass, depend on the solid, whereas the mass and charge of planets or electrons do not depend on their environment. Quasi-particles emerge from collective excitations. This gives rise to an effective mass which is not fixed, but depends on the internal interactions of the solid as well as on external fields.

Let us compare them with physical particles such as electrons, photons, etc. Quasi-particles behave like quantum mechanical particles or non-relativistic field quanta. Some of them have the electron charge and spin, while others have charge and spin zero, and yet others are even bound systems of excitations which form quasi-atoms. Many quasi-particles may occur in approximately localized states. In measurements of such states, they are found to be approximately localized in accordance with Heisenberg's uncertainty relation, which puts limitations on the momentum width Δk and the spatial region Δr.

Further particle-like features add to the effective mass, momentum, charge, and spin which quasi-particles may carry. In several respects, not only do they behave independently of the rest of the solid, but they also behave approximately independently of each other, like non-interacting or uncoupled quantum particles. Their wave functions superpose in such a way that they give approximately linear contributions to the energy of a solid. They behave like fermions or bosons, and interact like physical particles.

Since quasi-particles only behave *approximately* in this way, their interactions have to be taken into account in a *second* step. They give rise to scattering and polarization effects. For example, a charged quasi-particle may lose a quantum k of its momentum to core vibrations, or vice versa. Such an energy transfer is identical to the emission or absorption of a phonon. It can be described through the corresponding creation and annihilation operators q_k^+, q_k. There are many possible interactions of this kind. They are calculated with the tools of quantum mechanical scattering theory, in terms of the scattering process of physical particles and by means of perturbation theory.

In all these respects, they are like physical particles with an effective mass, energy, charge, etc. They are described theoretically in terms of real field quanta in accordance with (QP$_R$). They can be prepared in accordance with the operational wave concept (OW) and, under certain conditions, detected in accordance with the operational particle concept (OP). They obey conservation laws and the corresponding sum rules for dynamic properties. Hence if quasi-particles are fake entities, then so are all kinds of subatomic particles, too. The free electronic charge carriers, phonons, or excitons in a solid are as spooky or unspooky as free electrons or protons in the scattering experiments of particle physics, or the photons and atoms in the *which-way* experiments of quantum optics. Their only distinguishing mark is that they emerge in solids, on which their effective dynamic properties depend.

12.3.4 Comparison with Virtual Particles

The analogy between the theory of quasi-particles and quantum field theory goes further than explained in Sect. 12.3.1. The calculation of perturbation series of interactions is analogous to the calculation of the S-matrix elements of quantum field theory, including the use of Green's functions, the calculation of Feynman diagrams, and the superposition of processes in which virtual quasi-particles are involved. In addition, there is an analogy with the non-empty vacuum of quantum field theory and its effects on the charge and mass of the field quanta, concerning a feedback between the interactions within a solid and its ground state. There are always interactions between the vibrations of the ionic cores and the electrons of the valence band of a solid which *cannot* be taken into account using the approximation methods sketched above. They affect the difference between the lowest excited states and the ground state of a solid as follows:

> Thus a great deal of the electron–phonon interaction must be thought of as already included in the definition of the phonon excitation itself, just as the electrons which are being scattered are not really single independent electrons, but quasi-particles.

> The problem is remarkably like that of renormalization in field theory; the 'physical' quasi-particles and phonons we see are not the same at all as the 'bare' particles we can simply think about, and their experimental properties—energy, interactions, etc.—include contributions from the cloud of disturbances surrounding the bare particles.[26]

This gives rise to effects which mean that the "preliminary definition" of the "elementary excitation" no longer holds. Due to the many possible scattering processes, the quasi-particle states are not *really* eigenstates of the Hamiltonian of the whole system.[27]

At this point, it is natural to compare the quasi-particles of condensed matter physics with the virtual field quanta of quantum field theory. Are they as fictitious as the virtual particles of a single Feynman diagram in quantum field theory? Obviously *not*. In contradistinction to the former, they *come on their own as quantized appearances*, even though they are due to collective excitations. In particular, they may be prepared and detected in experiments. Hence it is not only through the mathematical approximation methods of condensed matter physics that they enter the stage. In addition, their calculation explains details of the energy band structure of a solid which the Hartree–Fock approach cannot explain; and it predicts details concerning the charge structure of a solid with impurities and imperfections. All this has been investigated in experiments. And it is most successful in predicting the phenomenological macroscopic behavior of solids, and in particular magnetic states or superconductivity.

---

[26] Anderson 1997, p. 116. In the following, Anderson mentions that the analogy ends at the divergences of quantum field theory, which fortunately do not occur when calculating the interactions of quasi-particles.

[27] One way of dealing with this problem is the "full renormalization" approach, which is once again similar to renormalization in quantum field theory; see Anderson 1997, p. 120.

Insofar as quasi-particles are localized in a crystal, they exist in the same sense as the spatial structure of subatomic scattering centers in the domain of non-relativistic scattering energy. They contribute to the quasi-classical charge distribution within a solid in the same sense as the squared amplitude of the many-particle wave function of atoms gives rise to a quasi-classical charge distribution.[28] This is particularly true of the unsharply localized wave packets of charged quasi-particles or of phonons which clump around an impurity, making a localized target for the scattering of charged quasi-particles.

According to the analogy with quantum field theory, quasi-particles are onto-logically equivalent to *real* field quanta of some given observable mass or energy. They are *real energy states* which underlie energy conservation, in contradistinction to virtual particles which may be 'off shell', i.e., violate energy conservation during the interaction. Otherwise, it would make no sense at all to define *virtual* quasi-particles by analogy with the virtual particles occurring in the perturbation expansion of the *S*-matrix of quantum field theory, a definition which does indeed complete the analogy between real field quanta and quasi-particles.

Hence, quasi-particles and virtual particles have in common only that, in con-tradistinction to physical particles such as free electrons, they do not come on their own. They belong to collective effects. The relation between the particles and the collective effect, however, is the other way round in the two cases. Virtual particles cannot be measured, but their theory tells us that they give rise to measurable collective effects. In contrast, quasi-particles can be measured, and the theory tells us that they are due to collective effects. With regard to the operational content of the concept, quasi-particles are not to be compared with virtual particles, but rather with the Lamb shift, the Casimir effect, etc., that is, to the observable phenomena of physics, and not their unobservable causes.

12.3.5 Comparison with Matter Constituents

According to the conservation laws of physics, the quasi-particles in solids do indeed contribute to the constitution of matter. They underlie the same kind of conservation laws and sum rules as the constituent models of atomic, nuclear, and particle physics. The quark model tells how the proton and neutron are made up of quarks and gluons, respecting the conservation laws and sum rules of energy, momentum, spin, and charges. The scattering experiments of high energy physics show that, in addition to the so-called (real) 'valence' quarks, a 'sea' of virtual quark–antiquark pairs and gluons contribute to the momentum and spin of the proton and neutron. Hence, according to the quark model, collective effects pro-duced by virtual particles contribute to the constitution of protons and neutrons in the atomic nucleus. Nuclear physics tells how the atomic nucleus is made up of

---

[28] See Falkenburg 2007, Chap. 4.

protons and neutrons, respecting the sum rule for mass, binding energy, and the mass defect which derives from relativistic mass–energy conservation. Atomic and molecular physics tell us how atoms are made up of the nucleus and the electron shell, how the atoms combine to form molecules, etc. And condensed matter physics explains how a transistor is made up of semi-conducting layers and the quasi-particles in the corresponding bands, in accordance with sum rules based on charge conservation.

12.4 Back to Scientific Realism

After all, quasi-particles compare to physical particles, matter constituents, and the measurable phenomena explained by virtual particles rather than the unobservable causes of the phenomena. If quasi-particles are fake entities, physical reality altogether is made up of fake entities.

Nevertheless, in the article "Manipulative Success and the Unreal" of 2003, Gelfert argued that quasi-particles, and in particular the holes in a semiconductor, are fake entities which counter Hacking's famous reality criterion "If you can spray them, they exist".[29] Gelfert's paper aims to demonstrate by means of counterexamples that Hacking's criterion of manipulative success is not a sufficient condition for the existence of an entity. To his counterexamples belong holes as a prototype of quasi-particles. Without discussing his line of reasoning in detail, I want to show that the way he employs quasi-particles as a counterexample against Hacking is at odds with the properties of quasi-particles.

According to Gelfert, it is uncontroversial that quasi-particles can be manipulated in experimental practice. They may be used as markers in crystals. But at least holes seem to be absences rather than genuine entities. If they can be manipulated even though they do not exist, however, Hacking's reality criterion seems to be refuted:

> Holes are at best fake entities—though 'sprayable' ones .... For 'manipulation' to be a success term, entities must be truth-makers for positive existential statements. Absences— and a whole class of more complex quasi-entities ...—do not satisfy this condition.[30]

Here, he claims that holes are only one striking case of quasi-particles, which he calls quasi-entities. He considers them not to be genuine entities, even though they may be used with manipulative success and hence in his view counter Hacking's reality criterion. Let me split my discussion of this claim into two different issues: Are holes fake entities? And, what about quasi-particles in general? Are there any reasons to consider them to be fake entities, in spite of the above results that they compare to real rather than virtual field quanta?

---

[29] Gelfert 2003, against Hacking 1983, pp. 22–25.
[30] Gelfert 2003, p. 255.

12.4.1 Are Holes Fake Entities?

Apart from the electron holes in a semiconductor, there are many kinds of holes in the world: the holes in a pair of socks that have been worn for a long time, the holes in a piece of Emmenthal cheese, the air pockets which may jolt an aircraft during its flight, the black hole in the center of the Milky Way, and so on. No one would say that the holes in a given pair of socks or the holes in the Emmenthal cheese do not exist, or (to put it in philosophical terms) cannot be the truth-makers for positive existential statements. My son's socks either have holes or they don't. If they have, they have another topological structure than if they don't. Both kinds of socks, or their topological structure, are likewise truth-makers for existential statements about holes or their absence. (By the way, is the claim 'There is no hole' a positive or a negative existential statement??) The same holds for the holes in the Emmenthal cheese, which are physical air bubbles in a piece of cheese, or an air pocket, that is, a zone of lower air pressure in which an aircraft drops abruptly from its original flight height. All these kinds of holes are directly observable. No one would claim that they do not exist, even though they are absences, namely a local lack of knitwear, cheese, or air pressure. Admittedly, they do not disprove Gelfert's way of reasoning, since they are not usually used with manipulative success in order to perform some kind of experiment.

The black holes of astrophysics and cosmology are different. They are not absences, but singularities of spacetime. Since no light or matter inside the Schwarzschild radius can escape them, they are unobservable in principle. Nevertheless, they are inferred from observable astrophysical phenomena such as stars rotating with very high speed around them, and they give rise to the prediction of Hawking radiation, a quantum phenomenon. They, too, are the subject of positive existential statements, which are the truth-makers of their theory. However, given that cosmological objects are not subject to manipulation in terrestrial experiments, they do not disprove Gelfert's way of reasoning either.

Even though none of these kinds of holes seems to be like the hole in a semiconductor, the examples show that Gelfert's crucial premise is far from being plausible. If a hole is used with manipulative success in an experiment, then the fact that it is a hole, and hence an absence of something in the world, does not imply that it is a fake entity. And so it is for the hole in a semiconductor, that is, a missing electron in a lattice of subatomic particles. The lattice as a whole may be considered to be a carrier of the positive charge, but the missing electron seems not to be an entity but rather a genuine absence, which is due to a kind of ionization process. Nevertheless, as shown in the preceding sections, the hole in a semiconductor has all kinds of effective particle properties. In particular, it may be localized under certain conditions, i.e., detected in the sense of the operational particle concept (OP). For this very reason, even a fool-blooded empiricist should accept its existence. And in this sense, the electron hole in a semiconductor is indeed the truth-maker of the positive existential statement "There is an electron hole of effective mass m_{eff} that has been measured." This claim may seem counterintuitive, but it is no more so than any existential claim about subatomic quantum particles and quantum waves in general. Indeed, the existence of the holes in a semiconductor does not counter Hacking's reality criterion at all. If holes can be prepared in such a way that they may be used as markers in a crystal, this manipulative success shows that their preparation is some kind of manipulation. And if the preparation and detection procedures of quantum physics are ontologically equivalent, the prepared quantum state of a hole exists and the manipulative success agrees perfectly with Hacking's reality criterion.

12.4.2 What About Quasi-Particles in General?

Gelfert claims that holes are only one particular class of fake (or quasi-) entities, amongst which he classifies all kinds of quasi-particles or collective effects in solids. Based on a discussion of spin waves and excitons, he argues that they are not genuine entities even though they have real causal powers, behave like particles, and hence can be used with manipulative success in experiments (Gelfert 2003, pp. 256–260). In order to criticize Hacking's reality criterion, he makes an epistemological and an ontological claim. The epistemological claim concerns Hacking's point that entity realism is more robust if it is based on the experimenter's background knowledge or familiar "home truths" rather than on highly sophisticated theoretical knowledge. According to Gelfert, the origin of the causal powers of quasi-particles is opaque to the experimenter, given that the theory of collective excitations is complicated. Indeed, the theory of quasi-particles is based on the quantum theory of a many-particle system and the formal apparatus of quantum field theory. (The question of whether the occupation number formalism is too complicated for experimenters is debatable, but the apparatus of perturbation theory including the tools of renormalization is indeed far from being simple.) Hence, in the case of quasi-particles, the manipulative success is combined with lack of knowledge about what kind of entity the experimenter is dealing with:

> The properties attributed to quasi-particles, such as the effective mass of electron holes or the lifetime of excitons, turn out to be functions of the total many-body system. It is important to note that whereas the properties of quasi-particles can be exploited experimentally, the composite nature of their collective dependency is opaque to experimental methods. The experimenter who sprays a quasi-particle is, in a sense, blind not to manipulative success generally but to the nature of what he uses as a tool.

Hence, even if quasi-particles can be used with manipulative success, this matter of fact does not justify any specific ontological inference to the nature of this tool. This is obviously true. However, we saw that the manipulative success is based on the preparation of the quantum state of a quasi-particle, which is defined in terms of measurable properties such as mass/energy and charge. Therefore, even if it is true that "the composite nature of their collective dependency is opaque to experimental methods", the ontological structure of quasi-particles is well-known, given that it is defined in terms of the effective dynamic properties of the quantum state that has been prepared. In this regard, the experimenters are *not* blind to to the nature of their tools. They know precisely that they are dealing with a quantum state of well-defined mass/energy, spin, and charge. And we have seen before that, according to a modest ontological approach, particles are no more than bundles of dynamic properties (mass/energy, spin, parity, and various kinds of charges) which underlie conservation laws and obey the corresponding symmetries (see the discussion of the symmetry particle concept (SG) in Sect. 12.2).

These considerations, however, also undermine Gelfert's ontological claim that quasi-particles are collective effects rather than genuine entities. In his view, the nature of quasi-particles as collective effects in a solid is at odds with the familiar home truths about what a genuine entity should be like:

> If one were to grant quasi-particles the same degree of reality as electrons, one would violate the very intuitions that lie at the heart of entity realism, namely, that there is a set of basic substantive entities that have priority over composite or derivative phenomena.[31]

> ... even on the non-theoretical body of home truths that the entity realist is forced to admit, we know that solids consist of crystals formed by atoms and of electrons travelling through the crystal, rather than of a plethora of emergent quasi-particles. This knowledge does not involve a theoretical inference; rather it represents a preservation of home truths.[32]

Based on my above line of reasoning, I reject both ontological claims. First, the "very intuitions that lie at the heart of entity realism" are already at odds with the rise of quantum theory and the failed attempts to establish a convincing version of local realism mentioned in Sect. 12.1. Second, due to the conservation laws and sum rules of subatomic physics there are good reasons to range the quasi-particles among the matter constituents. I can see no reasons that counter the claim that quasi-particles have the effective properties of physical particles, and hence contribute to the mass, charge, etc., of the solid.

As far as I can see, Gelfert's conclusion that quasi-particles are not genuine entities, and (given their manipulative success) thus counter Hacking's reality criterion, is based on an ambiguity in his notion of "home truth". One should carefully distinguish the physicist's and the philosopher's "home truths". The physicist's "home truths" consist in familiar background knowledge about measurement methods, physical quantities such as mass/energy, charge, spin, etc., and also (a century after the rise of quantum theory) basic knowledge about quantum mechanics, including the preparation of quantum states which guarantees manipulative success in experiments with quasi-particles. In contradistinction to this background knowledge of quantum physics, however, the philosopher's "home truths" about what genuine entities should be like are rooted in metaphysics. They are based on traditional metaphysical ideas about independent substances, which are at odds with the structure of quantum theory. Gelfert seems to adhere to such traditional metaphysical ideas, since he argues as follows:

---

[31] Gelfert 2003, p. 257.
[32] Gelfert 2003, p. 259.

The entity realist, thus, is faced with the dilemma of either following the criterion of manipulative success and ending up with a permissive form of inflationary realism that violates basic home truths, or trying to build on these home truths while depriving himself of the very theoretical tools necessary for "explaining away" quasi-particles as collective many-body excitations....

In short, if the entity realist were to bite the bullet on quasi-particles, he would have to give up either on entities as we know them, or on being a realist. Neither seems to be a viable option.[33]

In my view, the main metaphysical lesson taught by quantum mechanics and quantum field theory is that, in the quantum domain, the kind of entities as we know them from classical physics and traditional metaphysics do not exist. Hence, if the scientific realist is willing to believe in the existence of *any* quantum particles, such as electrons, protons, and photons, she must indeed bite the bullet when it comes to quasi-particles, too.

12.5 How Do Quasi-Particles Exist?

Let me now draw my general conclusions about the ontological status of quasi-particles. As shown in Sect. 12.3, they are *real* collective dynamic effects of the constituents of a solid. As quantized energy states which obey conservation laws and sum rules, they are not fake entities. They *exist* in the same sense as electrons, protons, and photons do, namely as bundles of dynamic properties (mass/energy, spin, charge, etc.) which underlie conservation laws and obey the corresponding symmetries. Indeed, they can be prepared and detected like physical particles. I have argued that for this reason they compare to real rather than virtual field quanta, given the formal analogy between the theory of quasi-particles and quantum field theory. I conclude that they are ontologically on a par with free electrons, protons, neutrons, as well as the subatomic matter constituents of atomic, nuclear, and particle physics.

Quasi-particles do *not* of course exist as *absolutely* independent entities, such as the classical particles were supposed to be. But this is also true of electrons, photons, and the corresponding spooky waves propagating through the branches of a *which-way* experiment. Quasi-particles exist as *approximately* independent collections of energy, mass, charge, and spin, which are approximately independent of each other, and even appear to be independent of the solid in which they emerge. Due to the mathematical problems of the many-particle quantum mechanics of a complex system, the quantitative prediction of their precise excitation energy may be bad. But this does not affect their existence as discrete energy states of a solid which may sometimes even be localized.

Indeed, quasi-particles are as real as a share value at the stock exchange. The share value is also due to a collective effect (as the very term indicates), namely the

---

[33] Gelfert 2003, p. 259.

collective behavior of all investors. The analogy may be extended. It is also possible to 'spray' the share value in Hacking's sense, that is, to manipulate its quotation by purchase or sale for purposes of speculation. Its free fall can make an economy crash, its dramatic rise may make some markets flourish. And the crash as well as the flourishing may be local, i.e., they may only affect some local markets. But would we conclude that the share value does *not exist*, on the sole grounds that it is a *collective effect*? Obviously, share values as well as quasi-particles have another ontological status than, say, Pegasus. Pegasus does not exist in the real world but only in the tales of ancient mythology. But quasi-particles exist in real crystals, as share values exist in real economies and markets. Both concepts have a well-defined operational meaning, even though their *cause cannot* be singled out by experiments or econometric studies. But the same is unfortunately true of the cause of an electron track. The track is not continuous but due to a series of subsequent position measurements which are *not* caused by an individual, classical particle.[34]

Therefore, the alleged (philosophical!) "home truths" of entity realists about real entities are slippery and delusive, given that they are rooted in traditional metaphysics. The familiar philosophical "home truths" about what genuine entities should be like are in apparent conflict with the "home truths" of current quantum physics. If it is a philosophical home truth that electrons exist as entities in their own right (i.e., in the sense of substances-on-their-own), and if this may be taken as a case against the existence of quasi-particles, all the worse for the philosopher's metaphysical home truths.

References

Anderson, P.W.: Concepts in Solids. World Scientific, Singapore (1997)

Bohm, D.: A suggested interpretation of the quantum theory in terms of hidden variables, I and II. Phys. Rev. **85**, 166179, 180. Repr. (In: Wheeler, J.A., Zurek, W.H. (eds.) Quantum Theory and Measurement. Princeton University Press 1983) (1952)

Cartwright, N.: How the Laws of Physics Lie. Clarendon Press, Oxford (1983)

Cartwright, N.: Natures Capacities and their Measurement. Clarendon Press, Oxford (1989)

Ehrenfest, P.T.: Begriffliche Grundlagen der Statistischen Auffassung der Mechanik. Encycl. Math. Wiss. **4**, Art. 32 (1911)

Falkenburg, B.: Particle Metaphysics. A Critical Account of Subatomic Reality. Springer, Heidelberg (2007)

Falkenburg, B.: 2012: Was sind subatomare Teilchen? In: Esfeld, M. (ed.) Philosophie der Physik, pp. 158–184. Suhrkamp, Frankfurt am Main (2012)

Gelfert, A.: Manipulative success and the unreal. Int. Stud. Philos. Sci. **17**(3), 245–263 (2003)

Hacking, I.: Representing and Intervening. Cambridge University Press, Cambridge (1983)

Kuhn, T. W.: The Structure of Scientific Revolution, 2nd edn. University of Chicago Press, Chicago, 1970 (1962)

---

[34] Van Fraasen 1980, pp. 58–60, interprets this example as a case for empiricism, an interpretation that can be challenged, however; see my discussion in Falkenburg 2007, Sect. 2.6.

Mach, E.: Die Mechanik in ihrer Entwicklung historisch-kritisch dargestellt. Leipzig: Brockhaus (1st edn) Quoted after the repr.: Darmstadt: Wissenschaftliche Buchgesellschaft (1991, repr. of the 9th ed.: Leipzig: Brockhaus 1933). Engl. Transl.: The Science of Mechanics. A Critical and Historical Account of its Development. La Salle, Ill: Open Court 1960 (6th edn.) (1883)

Maxwell, G.: The ontological status of theoretical entities. In: H. Feigl, Maxwell, G. (eds.) Scientific Explanation, Space, and Time, pp. 3–27. (Minnesota Stud Philos Sci 3). University of Minnesota Press, Minneapolis (1962)

Maxwell, J.C.: Letter to Stokes. In: Harman, P.M. (ed.) The scientific letters and papers of James Clerk Maxwell, vol. I, 1846–1862, pp. 606–609. Cambridge University Press, Cambridge (1859)

Planck, M.: Die Einheit des physikalischen Weltbildes. In: *Vorträge und Erinnerungen.* Darmstadt: Wissenschaftliche Buchgesellschaft, pp. 28–51 (9th edn., 1965) (1908)

Psillos, St: Scientific Realism. How Science Tracks Truth. Routledge, London (1999)

Scheibe, E.: Between Rationalism and Empiricism: Selected Papers in The Philosophy of Physics (ed. by B. Falkenburg). Springer, New York (2001)

Scheibe, E.: Die Philosophie der Physiker, 2nd edn. C.H. Beck, München (2007)

van Fraassen, B.C.: The Scientific Image. Clarendon Press, Oxford (1980)

Wigner, E.P.: On unitary representations of the inhomogeneous lorentz group. Ann. Math. **40**, 149–204 (1939)

Worrall, J.: Structural realism: the best of both worlds? Dialectica **43**, 99–124. (Repr. In: Papineau, D. (ed.) The Philosophy of Science, p. 139–165. Oxford University Press 1996) (1989)

Chapter 13
A Mechanistic Reading of Quantum Laser Theory

Meinard Kuhlmann

13.1 Introduction

I want to show that the quantum theory of laser radiation provides a good example of a mechanistic explanation in a quantum physical setting. Although the physical concepts and analytical strategies I will outline in the following do admittedly go somewhat beyond high school knowledge, I think it worth going some way into the state-of-the-art treatment of the laser, rather than remaining at a superficial pictorial level. In the course of the ensuing exposition of laser theory, I want to show that the basic equations and the methods for solving them can, despite their initially inaccessible appearance, be closely matched to mechanistic ideas at every stage.

In the quantum theory of laser radiation, we have a decomposition into components with clearly defined properties that interact in specific ways. This dynamically produces an organization that gives rise to the macroscopic behavior we want to explain. I want to argue that a mechanistic reading is not one that can be overlaid on laser theory so that it coheres with the mechanistic program, but rather that the quantum theory of the laser is inherently mechanistic, provided that the notion of a mechanism is slightly broadened. As I will show, the pieces required to identify the workings of a mechanism can be seen directly on the level of the basic equations. And this applies even more clearly to the following derivation than to the more picturesque semiclassical derivations, because it starts on the most basic level of quantum field theory, where *all* the relevant parts of the laser mechanism are described in detail, e.g., atoms with internal structure and specific behavior in isolation and interaction.

M. Kuhlmann (✉)
Philosophisches Seminar, Johannes Gutenberg University,
55099 Mainz, Germany
e-mail: mkuhlmann@uni-mainz.de

© Springer-Verlag Berlin Heidelberg 2015
B. Falkenburg and M. Morrison (eds.), *Why More Is Different*,
The Frontiers Collection, DOI 10.1007/978-3-662-43911-1_13

When all is said and done, the quantum theory of laser radiation is a neat example of a *mechanistic* explanation because we have an explanation that shows how the stable behavior of a compound system reliably arises purely on the basis of interactions between its constituents, without any coordinating external force.[1] The proof that quantum laser theory can be understood as supplying mechanistic explanations has a number of important implications. Most importantly, it shows that mechanistic explanations are not limited to the classical realm. Even in a genuinely quantum context, mechanistic reasoning can survive.[2] Mechanistic explanations are attractive because they often provide the best route to effective interventions. Moreover, understanding the general mechanisms involved in self-organizing systems such as the laser allows one to transfer certain results to other less well understood systems where similar mechanisms (may) operate.

13.2 What Is a Mechanism?

According to the mechanical philosophy of the seventeenth century, one could explain everything by the mechanical interaction (push and pull) of tiny building blocks. In contrast, today's mechanists, also called the "new mechanists", have a more modest point. They do not claim that everything can and must be explained in terms of mechanisms. For instance, electromagnetic interactions may well not be mechanistically explicable. The crucial point for the new mechanists is that mechanistic explanations play the dominant role in most sciences, something not appropriately represented in the standard philosophy of science. In many cases, biology, but also physics, and in particular in its applied branches, do not focus primarily on laws. They still play a role, but not a prominent one. Accordingly, the philosophy of science should be amended as far as mechanisms are concerned.

Whereas the *interaction theory of mechanisms* (Glennan 2002) says that a "mechanism for a behavior is a complex system [in the sense of compound system, MK] that produces that behavior by the interaction of a number of parts, where the interactions between parts can be characterized by direct, invariant, change-relating generalisations" (p. S344), the *dualistic approach* (Machamer et al. 2000) has it that "[m]echanisms are entities and activities organised such that they are productive of regular changes from start or set-up to finish or termination conditions" (p. 3). Today there are a number of proposals for a consensus formulation. Illari and Williamson (2012) propose the following unifying characterization of mechanisms: "A mechanism for a phenomenon consists of entities and activities organized in such a way that they are responsible for the phenomenon" (p. 120). Thus Illari and Williamson see the identification of three elements as essential for a mechanism, namely (i) responsibility for the phenomenon, (ii) "entities and activities", and (iii)

---

[1] This fact is also the reason why the laser is a paradigmatic example of a self-organizing system.

[2] See Kuhlmann and Glennan (2014) for a more comprehensive discussion of why quantum mechanics seems to undermine mechanistic explanations, and why in fact it doesn't.

organization. Bechtel and Abrahamsen (2011) see as the "key elements of a basic mechanistic explanation [...] (1) the identification of the working parts of the mechanism, (2) the determination of the operations they perform, and (3) an account of how the parts and operations are organized so that, under specific contextual conditions, the mechanism realizes the phenomenon of interest" (p. 258). I will take this proposal as the background for the following analysis, in which I will check whether and how these requirements are fulfilled.

13.3 Quantum Laser Theory Read Mechanistically

13.3.1 The Explanandum

In his famous paper on black body radiation Einstein (1917) introduced the idea of 'induced' or 'stimulated' emission of light quanta, later called photons. This idea already suggests the possibility of amplifying light of a chosen wavelength in a systematic way, and this is what was realized technically by the laser (light amplification by stimulated emission of radiation) in 1960. Lasers are light sources with outstanding properties, such as very high monochromaticity (temporal coherence), a high degree of collimation (spatial coherence), and high intensity of radiation. For weak energy supply, lasers radiate conventional lamp light, e.g., a superposition of numerous wavelengths. Once the energy supply exceeds the so-called laser threshold, all the atoms or molecules inside a laser begin to oscillate in a single common mode, emitting light of (ideally) just one wavelength and therefore one color (Fig. 13.1).

The aim of laser theory is to explain how the interaction of the photon-emitting atoms produces laser light. That is, the goal is to calculate the dynamics of the compound system, i.e., the laser, in terms of its interacting subunits. The dynamics is described by differential equations, i.e., by equations that contain a function together with its derivative(s). Differential equations constitute a core part of every physical theory. With a differential equation which relates a state function to its temporal derivatives, knowing the state at one time allows one to determine the state at all later times. For a complex system, such as a laser, the basic differential equations can be horrendously complicated because of their sheer number and mutual coupling. For example, if the number of laser-active atoms is of the order of 10^{14}, one gets 10^{18} differential equations. Thus, apart from determining the relevant set of differential equations, the ambition of laser theory is to solve this system of differential equations, which is a formidable task.

13.3.2 Specifying the Internal Dynamics

In the semiclassical laser theory, only the atoms are described by quantum theory, whereas the electrical field in the laser cavity is assumed to be classical. This is a

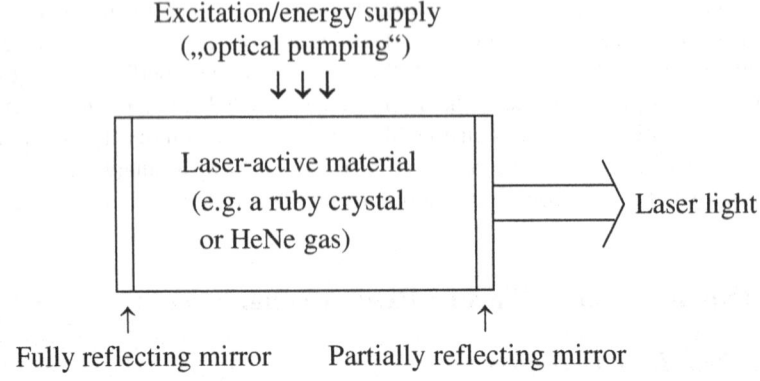

Fig. 13.1 Schematic design of a laser

comparatively simple but already very powerful approach. However, in the present context, I start with the more advanced quantum theory of laser radiation for an obvious reason: we should avoid assuming any classical physics when we want to show that classical mechanistic concepts are applicable even in the quantum realm. Moreover, from an ontological point of view, the quantum theory of laser radiation has yet another advantage: the basic equations for the dynamics of laser radiation can be derived from first principles, i.e., by starting with fundamental equations. In the following I will introduce these basic dynamical equations in some detail, because they are crucial for demonstrating the mechanistic nature of laser theory.

The first important set of equations for laser radiation are the **field equations**, which specify the time dependence of the electromagnetic field operators b_λ. In the classical case, the b_λ are the amplitudes of possible states of oscillation of the electromagnetic field inside the laser cavity or 'resonator', counted by the index λ (the wavelength). This means that each individual $b_\lambda(t)$ specifies how much the λ-th mode is excited. In a quantum setting, b_λ^+ (the complex conjugate of b_λ) and b_λ become creation and annihilation operators for the laser field, i.e., each occurrence of b_λ^+, or b_λ, in a formula (e.g., a Hamiltonian, see below) represents the creation, or annihilation, of a photon with quantum number λ (classically the wavelength).

The basic equations in laser theory capture the dynamics of the essential quantities. In the quantum setting used in laser theory, the time dependence of an operator A is determined by the Heisenberg equation of motion[3]

---

[3] I work in the so-called 'Heisenberg picture'. As is well-known quantum mechanics can be formulated in different mathematically and physically equivalent ways. The two best-known representations or 'pictures' are the Schrödinger picture and the Heisenberg picture. Quantum mechanics is mostly formulated in the Schrödinger picture, where the state is time-dependent while the observables for position and momentum are time-independent. In the Heisenberg picture, on the other hand, observables carry the time-dependence, whereas the states are time-independent. Mathematically, the Heisenberg picture is related to the Schrödinger picture by a mere basis change, and thus physically both pictures lead to the same measurable quantities, of course. In

$$\frac{d}{dt}A \equiv \dot{A} = \frac{i}{\hbar}[H, A] \equiv \frac{i}{\hbar}(HA - AH), \qquad (13.1)$$

where H denotes the Hamilton operator, or 'Hamiltonian' for short, which represents the total energy of the system.[4] So the first step in the quantum theory of laser radiation—as in most other quantum physical treatments of the dynamics of a given system—consists in finding the Hamiltonian of the system. In our case, the Hamiltonian of the whole system, i.e., the laser, can be decomposed as follows

$$H = H_f + H_A + H_{Af} + H_{B_1} + H_{B_1-f} + H_{B_2} + H_{B_2-A} \qquad (13.2)$$

where H_f denotes the Hamiltonian of the light field, H_A that of the atoms, and H_{Af} that of the interaction between the atoms and the field; H_{Bi} is the Hamiltonian of heat bath i and H_{Bi-f} of the interaction between heat bath i and the field. Even on this very first level, mechanistic ideas can already be clearly identified, since the total Hamiltonian is neatly split up into parts that comprise the behavior of the system's components in isolation, followed by all the interactions between these components and with any other relevant systems. A more detailed description of all these Hamiltonians will be given now.

H_f is the Hamiltonian for the electromagnetic light field, and H_A the Hamiltonian for the *atoms* inside the laser, which in turn sums over the Hamiltonians of all the individual atoms, i.e., each atom has its own Hamiltonian—notwithstanding the indistinguishability of "identical" quantum particles.[5] These two parts of the total Hamiltonian determine the behavior of the light field and of the laser atoms in isolation, i.e., if there is no interaction whatsoever between the field and the atoms or with any other entities, such as the environment of the system. The next part of the total Hamiltonian, H_{Af}, captures the way the atoms interact with the field modes. One term that appears in this Hamiltonian for the interaction between the field modes and the atoms is $a_{1,\mu}^+ a_{2,\mu} b_\lambda^+$, which represents (read the formula from right to left) the creation of a photon in field mode λ, the annihilation of an electron in state 2 (the higher energy level), and the creation of an electron in state 1 (the lower energy level), a sequence which can be grasped quite intuitively (see Fig. 13.2).

So far we have three essential parts of the total Hamiltonian, which seems to be all we need to know in order to determine the dynamics that leads to laser light. And in fact, the semiclassical laser theory gets pretty far without considering

---

(Footnote 3 continued)
some respects, the Heisenberg picture is more natural than the Schrödinger picture (in particular, for relativistic theories) since it is somewhat odd to treat the position operator, for instance, as time-independent. Moreover, the Heisenberg picture is formally closer to classical mechanics than the Schrödinger picture. For this reason it is advantageous to use the Heisenberg picture if one intends to compare the quantum and the classical case, which I want to do for the laser.

[4] Note that the first and the last part of the above row of equations are just definitions, indicated by "$\equiv$".

[5] See Haken (1985), p. 236ff.

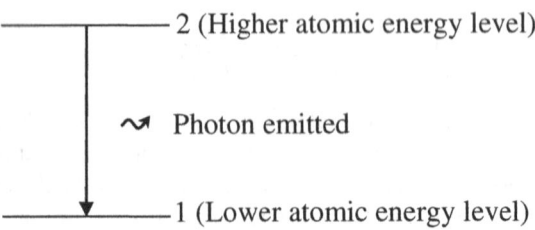

Fig. 13.2 Schematic representation of a field-atom interaction

anything other than the field (described classically), the atoms, and their interaction. However, it turns out that the semiclassical laser theory is unable to explain the transition from conventional lamp light to laser light, which occurs at the 'laser threshold', and some details of the coherence properties of laser light. What has been left out so far is damping. The light field inside the laser is damped due to the transmissivity of the mirrors and other cavity losses, and the atoms are also damped by various processes (more below). Damping of a quantity always produces fluctuations, which in turn have important consequences. In the Hamiltonian, damping is accounted for in terms of an additional interaction (or coupling) of the light field with a 'heat bath', called B_1, which is taken to cover all the above-mentioned processes, as well as a further interaction of the atoms with a second heat bath, B_2, each of which has its own Hamiltonian.[6]

Let us take stock. In order to understand what happens in a laser we need to know how the basic physical quantities, in quantum theory represented by operators, evolve in time. Due to the Heisenberg equation of motion (1), we need the total Hamiltonian of the laser in order to determine the time evolution of any operator in which we are interested. Thus the Hamiltonian characterizes those specifics of our system that determine how it evolves in time, in particular when its parts interact with each other. As we have seen the basic dynamical setting of the quantum theory of laser radiation—given by the total Hamiltonian of the laser system—rests on a clear separation of different relevant components, whose behavior is described both in isolation and in mutual interaction. This observation will play a crucial role in our philosophical assessment concerning the mechanistic nature of the quantum theory of laser radiation.

Now let us begin to actually write down the equations of motion for the relevant physical quantities. In other words, we want to formulate those equations that tell us the dynamics of the important quantities, i.e., how they evolve in time. The first set of such equations determines the dynamics of the laser's light field in terms of its electromagnetic field operators b (we already described the significance of b, just before we introduced the Hamiltonian that determines its dynamics). In order to get

---

[6] The fluctuations comprise thermal and quantum fluctuations, giving rise to additional statistical correlations between the atoms and the field. Bakasov and Denardo (1992) show in some detail that there are some corrections due to the "internal quantum nature" of laser light, which they call "internal quantum fluctuations".

those equations, we need to insert the relevant parts of the total Hamiltonian into the Heisenberg equation of motion (1), which gives us the following set of differential equations for the quantized *field* modes:

$$\dot{b}_\lambda = -i\omega_\lambda b_\lambda + \kappa_\lambda b_\lambda - i\sum_\mu g^*_{\mu\lambda}\alpha_\mu + F_\lambda(t) \qquad (13.3)$$

The first term refers to the freely oscillating field and the second to the damping of the field mode due to the interaction between the field and the heat bath in the laser cavity. The third term accounts for the interaction between the field and the atoms, and the fourth is an operator that describes a fluctuating force. The index μ counts the atoms that are excited inside the laser and produce the light field. ω_λ is the frequency of the λ-th mode, the coupling constant $g_{\mu\lambda}$ specifies the interaction between the μ-th atom and the λ-th mode, and α_μ is the operator for the polarization of the μ-th two-level atom.[7] Classically, the (formally identical) terms $-ig^*_{\mu\lambda}\alpha_\mu$ reflect the way the mode amplitudes (i.e., the b_λ) change due to the oscillating atomic dipole moments. Mathematically, these terms lead to a coupling with the next set of differential equations for the dynamics of the atomic variables α_μ (more below). The $\alpha_\mu \equiv a^+_{1,\mu}a_{2,\mu}$ represent the annihilation of an electron in state 2 (the higher energy level), while an electron in state 1 (the lower energy level) is created. The above-mentioned damping of the light field inside the laser is captured by the damping constant (or 'relaxation speed') κ_λ. Since damping of a quantity produces fluctuations in its turn, one introduces the stochastic force $F_\lambda(t)$, which accounts for fluctuations due to any kind of dissipation (loss of directed energy, e.g., by friction or turbulence).

The next sets of differential equations, the **matter equations**, determine the dynamics of the laser-active atoms inside the laser cavity. The first group of equations

$$\dot{\alpha}_\mu = -(iv+\gamma)\alpha_\mu + i\sum_\lambda g_{\mu\lambda}d_\mu b_\lambda + \Gamma_\mu(t) \qquad (13.4)$$

with Hermitian conjugate

$$\dot{\alpha}^+_\mu = (iv-\gamma)\alpha^+_\mu - i\sum_\lambda g_{\mu\lambda}d_\mu b^+_\lambda + \Gamma_{\mu+}(t) \qquad (13.5)$$

---

[7] One can make a few simplifications (single laser mode, coupling constant independent of λ and μ) which ease the ensuing calculations. However, in the present context they are not helpful for a better understanding because they require further explanation and justification and widen the gap with realistic situations. For this reason I use the equations on p. 246 in Haken (1985), but without the simplifications introduced on p. 123, and that means with additional indices, which are still there on pp. 121ff.

specify how the atomic polarization changes in time. These equations are again coupled with those for the field operators b_λ above, since the field has an effect on the dynamics of the atoms. Still another coupling stems from the occurrence of the variable d_μ, which describes the **atomic inversion** $d_\mu \equiv (N_2 - N_1)_\mu$, i.e. the difference in occupation number of the energy levels (which are taken to be two for simplicity) that the laser-active atoms can be in. In the end, the temporal change in the atomic inversion is given by the differential equations

$$\dot{d}_\mu = \gamma_{\parallel}(d_0 - d_\mu) + 2i \sum_\lambda \left(g^*_{\mu\lambda}\alpha_\mu b^+_\lambda - g_{\mu\lambda}\alpha^+_\mu b_\lambda \right) + \Gamma_{d,\mu}(t), \qquad (13.6)$$

which is the second group of **matter equations**. $\Gamma_\mu(t)$ and $\Gamma_{d,\mu}(t)$ account for those fluctuations that are connected with the damping constants γ and $\gamma_{\parallel}$.

The damping constants in the four laser equations above refer to different kinds of damping processes, and this is in fact crucial for the solution of the full system of coupled non-linear differential equations for laser light. Classically, the damping constant (or 'relaxation speed') κ_λ captures the damping of the field amplitude b_λ in the resonator if there is no interaction between the field mode and the laser atoms, e.g., due to the transmissivity of the mirrors. To put it another way, κ_λ is the decay constant of mode λ without laser activity. The constant γ describes the damping of the freely oscillating atomic dipole due to the interaction of the atoms with their environment, and $\gamma_{\parallel}$ refers to the damping of the atomic inversion due to incoherent decay processes such as non-radiative transitions (e.g., by emitting energy in the form of lattice vibrations or, quantum physically, 'phonons') or spontaneous emission.

13.3.3 Finding the System Dynamics

Now the aim of laser theory is to solve the above system of coupled differential equations, but this is impossible using conventional methods of fundamental physics. The crucial starting point for tackling this task is the empirical fact that there is a hierarchy of time scales, or speeds, for the relevant processes.[8] The characteristic time scales for the dynamics of the field modes b_λ and of the inversion d_μ are much longer than the time scale for the dynamics of the atomic polarization α_μ. This fact can be expressed in terms of inequalities for the characteristic time scales, or alternatively for damping constants (the reciprocals of the time scales):

$$T_b \gg T_d \gg T_\alpha$$
$$\kappa_\lambda \ll \gamma_{\parallel} \ll \gamma. \qquad (13.7)$$

---

[8] Hillerbrand (2015), Sect. 13.3.2 of this book, discusses this separation of time scales in the more general context of scale separation and its impact for the feasibility of micro-reduction.

This means that the atomic polarization α_μ (connected to the T_α, or the damping constant γ) reaches its equilibrium value faster than d_μ, and d_μ in turn faster than b_λ, where this equilibrium value of α_μ is—due to the coupling of the differential equations through the non-linear terms—determined by the slower quantities b_λ and d_μ.

The hierarchy of process speeds has an extremely important consequence for the solution of our system of differential equations: certain slow quantities, the so-called *order parameters*, can be treated as constant in time in comparison to the much faster changes in other quantities. While the order parameter, here the field mode b, arises internally through the radiation of all the atoms, the *control parameter* can be adjusted or controlled externally, e.g., by energy supply. In the following description I will provisionally use the language of *synergetics*,[9] which I will scrutinize in the next section. Since the field modes have the longest time scale, one particular b_λ wins the competition and dominates the beat, so to speak. Consequently, there is only one basic mode in the resonator (i.e., symmetry breaking[10]) and one can drop the index λ in the differential equations (single mode case). The next step consists in the formal integration of the differential equations for α_μ:

$$\alpha_\mu(t) = \int_{-\infty}^{t} f_\mu(\tau) d_\mu(\tau) b(\tau) d\tau + \widehat{\Gamma}_\mu(t). \tag{13.8}$$

Note that this step does not yet get us very far since $d_\mu(t)$ and $b(t)$ are not given explicitly, but only implicitly determined by the above differential equations. $\widehat{\Gamma}_\mu(t)$ denotes the result of an integration, and for the following analysis it is not important to know it in any detail. The same applies to the term $f_\mu(t)$.

Mathematically, the following pivotal step is based on the hierarchy of time scales. Since the slower parameters $d_\mu(t)$ and $b(t)$ can be viewed as constant (in time), they can be pulled out of the integrand so that one gets

$$\alpha_\mu(t) = d_\mu(t) b(t) \int_{-\infty}^{t} f_\mu(\tau) d\tau + \widehat{\Gamma}_\mu(t), \tag{13.9}$$

where the integral is solvable in an elementary way. Put in the language of *synergetics* again, this so-called adiabatic approximation means that the atoms "follow

---

[9] In the 1970s Hermann Haken established the interdisciplinary approach of *synergetics* by transferring certain general insights that he had gained in his work on laser theory (see Haken 1983). Synergetics is one of a few very closely related theories of self-organization in open systems far from thermodynamic equilibrium.

[10] The predominance of one particular mode throughout the entire laser defines a ground state that no longer exhibits the symmetry of the underlying fundamental laws. These laws thus have a hidden symmetry that is no longer visible in the actual state of affairs, i.e., it is "spontaneously broken".

the commands" of the order parameter. The mathematical result of this crucial step is that d_μ is eliminated from the system of differential equations as an independent variable. In other words, d_μ is "enslaved" by $d_\mu(t)$ and $b(t)$. The following steps implement the same procedure for $d_\mu(t)$ and $b(t)$. The final result is *one* equation for *one* variable, namely b, the order parameter. In this way it is possible to solve the seemingly intractable system of differential equations for laser light dynamics. Physically, to cut things short, the resulting dominance of the variable b explains why we get laser light with its outstanding properties such as (almost) mono-chromaticity, i.e., light with a single pure colour. In the next section, I will spell out in detail why this procedure for explaining the onset of laser light does in fact give us a *mechanistic* explanation.

13.3.4 Why Quantum Laser Theory is a Mechanistic Theory

As promised at the outset, I intend to show that the quantum theory of laser light fulfills all the requirements for a mechanistic explanation. In order to have a clear standard of comparison I use, as introduced above, the characterization by Bechtel and Abrahamsen (2011, p. 258), according to which the core ingredients of a mechanistic explanation are "(1) the identification of the working parts of the mechanism, (2) the determination of the operations they perform, and (3) an account of how the parts and operations are organized so that, under specific contextual conditions, the mechanism realizes the phenomenon of interest". I will proceed in two steps. In this section I will show that a first survey of quantum laser theory allows us to identify all three ingredients of a mechanistic explanation. To this end I will commence by comparing quantum laser theory with its semiclassical predecessor, which will help us to identify the dynamical structures of the laser light mechanism in the quantum treatment. In the second step (Sect. 13.4), I will discuss, and dissolve, a number of worries that seem to undermine a mechanistic reading of quantum laser theory.

Strikingly, the laser equations in a full quantum physical treatment are formally almost identical with the basic equations of the semiclassical laser theory. They can be understood and solved in close analogy with the semiclassical case. Even if one describes everything in terms of quantum physics and includes all the complexity of the situation, the resulting behavior does not change fundamentally in many respects. It just involves a certain number of corrections. But what does this tell us? Despite their remarkably congruent results, it seems that semiclassical and quantum laser theory cannot be taken equally seriously. Semiclassical theories are generally considered to have a dubious status. If QM is true and universally valid, then semiclassical laser theory is, strictly speaking, simply wrong.[11] It is only a very helpful approximation (see Norton 2012), but not the true story. However, in the context of my investigation, I want to make the following claim: the fact that semiclassical laser theory gets so many things right only shows how much classical mechanistic modeling survives in the quantum mechanical explanation.[12]

Since the continuity from the semiclassical reasoning to the quantum treatment refers in particular to the essential interactive processes that produce laser light, this means that, insofar as semiclassical laser theory is mechanistic, so is quantum laser theory. To make this point there is no need to go as far as saying that (semi-) classical reasoning is indispensable for a full understanding of quantum phenomena.[13] Neither is it necessary to claim that purely quantum mechanical explanations are inferior to (semi-) classical explanation in at least some respects, in particular concerning the dynamical structure that is responsible for the phenomenon to be explained.[14] All that is needed in the context of my study is the fact that there are

---

[11] Moreover, in semiclassical laser theory, not everything is correct. For instance, below a certain threshold, lasers emit conventional lamp light. Semiclassical laser theory cannot accommodate this fact.

[12] Cartwright (1983) exploits this similarity in a different way. According to her reading, the quantum physical and the semiclassical approach offer two different theoretical treatments, while they tell the same causal story. And since we thus have different theoretical treatments of the same phenomenon, the success of these explanations yields no evidence in favour of a realistic interpretation of the respective theories. Morrison (1994) objects to Cartwright's claim that the fate of the theoretical treatments is a supposedly unique causal story, saying that it is *not* unique. A closer survey of laser theory reveals that "there are also a *variety of causal mechanisms* [my emphasis, MK] associated with damping and line broadening" (Morrison 1994). Consequently, one has to look for something else that the different approaches share. Morrison argues that *capacities*, as introduced in Cartwright (1989), may do the job. However, as she then shows, there is also an insurmountable obstacle for telling a unique causal story in terms of capacities, if one understands capacities as entities in their own right. Against such a Cartwrightian reification of capacities, Morrison argues that, if one wants to describe laser theory in terms of capacities, there is no way around characterizing them in relational terms. Eventually, this could give us a unique causal story, albeit without any additional ontological implications about capacities as entities in their own right. While I think that Morrison's reasoning is generally correct, I think there is an alternative to saying that capacities can only be characterized in relational terms. I claim that the causal story of laser light is best caught in terms of mechanisms. In the context of mechanisms, it is much more obvious that we don't need, and should not reify causal powers, because the crucial thing is the interactive, i.e., causal organization of the system's parts.

[13] This is what Batterman (2002) claims: "There are many aspects of the semiclassical limit of quantum mechanics that cannot be explained purely in quantum mechanical terms, though they are in some sense quantum mechanical" (p. 109). [...] "It is indeed remarkable how these quantum mechanical features require reference to classical properties for their full explanation. Once again, these features are all contained in the Schrodinger equation—at least in the asymptotics of its combined long-time and semiclassical limits—yet, their interpretation requires reference to classical mechanics" (p. 110).

[14] Bokulich (2008) refrains from some of the stronger claims by Batterman arguing that "one can take a structure to explain without taking that structure to exist, and one can maintain that even though there may be a purely quantum mechanical explanation for a phenomenon, that explanation —without reference to classical structures—is in some sense deficient" (p. 219); [...] semiclassical

structural similarities in the way the dynamics is modelled in the (semi-) classical approach on the one side and the quantum treatment on the other.

So how then are the three requirements for a mechanistic explanation met by quantum laser theory? In order to arrive quickly at a comprehensive picture, I begin with a very brief account, which will be defended in the next section. First, the working parts of the mechanism are the atoms and the field modes. Second, the operations they perform are specified by those parts of the differential equations that only refer to the variable whose dynamics is determined by the differential equation. Third and finally, the account of how these parts and operations are organized so that they produce the phenomenon of interest is given (or rather completed) by the coupling between the different variables in the system of differential equations, together with the crucial observation about the vastly different process speeds (the scale separation in Hillerbrand's terminology). The "specific contextual conditions" are the various specifications of the setup. In the following section I will discuss and dispel a number of objections that might be brought against this identification of the three key elements of a mechanistic explanation in quantum laser theory.

13.4 Potential Obstacles for a Mechanistic Reading

Quantum laser theory as presented above is the full quantum version of a complex systems explanation for a phenomenon concerning the light field which, under certain conditions, arises in a laser. This somewhat cumbersome formulation is meant to comprise all three elements in the explanation of laser light that could block a mechanistic reading. First, it treats the laser as a *complex* system; second, it is a *field* theoretic explanation; and, third, it rests on *quantum* theory with its various differences from classical mechanics. For each of these three potential obstacles to a mechanistic reading, I want to concentrate on that aspect that seems most relevant to me, where the second and the third points are connected.

13.4.1 Is "Enslavement" a Non-mechanistic Concept?

The first potentially problematic point in the above argumentation that quantum laser theory offers a mechanistic explanation of laser light is concerned with the fact that the laser is treated as a complex system.[15] More specifically, the enslavement principle, which I have, following Haken, provisionally employed in Sect. 13.3.3,

---

(Footnote 14 continued)

explanations are deeper than fully quantum mechanical explanations, *insofar as* they provide more information about the dynamical structure of the system in question than the quantum calculations do" (p. 232). However, in the present context even these weaker claims are not needed.

[15] In Kuhlmann (2011) I deal with the general question of whether complex systems explanations can be understood as mechanistic explanations.

could be incompatible with a mechanistic reading. Thus we need to discuss the exact explanatory and ontological status of this principle. The concept of enslavement generalizes the notion of the order parameter that was introduced with the Ginzburg-Landau theory of superconductivity in the 1950s.[16] The core idea is that the fast parameters are "enslaved" by one (or a few) slow 'order parameter(s)'. For the laser, the field mode b is the order parameter, i.e., the enslaving variable. The order parameter is a quantity that refers to the whole composite system and which arises by the joint action of the component parts. At the same time, the order parameter has, or seems to have, a feedback on what these parts do. Once a macroscopic mode has developed in the laser, the emission behaviour of the single atoms is—due to the broken symmetry—no longer as free as it was before. In synergetics, this fact is expressed by saying that the macroscopic mode dominates or "enslaves" all the component parts.

If this causal language is interpreted realistically it means that a higher-level entity has some kind of autonomous causal power. However, such strong conclusions don't seem to be sustained by the theory. For instance, in laser theory, talk of an order parameter that enslaves the behaviour of the component parts is an unwarranted causal description of a mathematical procedure, because there is no reason why it should represent a corresponding physical process. Arguably the most detailed critique of the far-reaching claims of synergetics concerning the ontological status of enslavement has been put forward by Stephan (1999), Chap. 18. He argues that the crucial significance of the order parameter in synergetics is merely a matter of description: only a descriptive thesis about the *compressibility of information* is warranted, namely that the system behaviour can be adequately described by one or a few order parameters without any need to specify the behaviour of all individual parts. However, this compressibility of information doesn't licence a *compressibility of causal factors*, i.e., the different and much stronger claim that the order parameter is a causal agent in its own right, which determines the behaviour of the system's parts. In more abstract terms, Stephan diagnoses a logical fallacy of the type post hoc, *ergo propter hoc*: the fact that focusing on the order parameters allows us to predict the behaviour of the system does not imply that the order parameter causally determines the system with all its parts. The implausibility of rating order parameters as causal factors becomes most obvious by looking at applications of synergetics in the social sciences: the work climate, Stephan says (p. 237), doesn't enslave the behavior of the clerks because the work climate doesn't do anything at all.

As Hillerbrand (2015), Sect. 13.3.2 of this book, puts it, the "methodology known as the 'slaving principle' [...] allows one to drastically simplify the micro-reductionist description". However, this doesn't imply that the employed order

---

[16] The Ginzburg-Landau theory was initially a phenomenological theory that analyzed the occurrence of superconducting phase transitions by general thermodynamic arguments without using a microscopic underpinning (as later supplied by the Bardeen-Cooper-Schrieffer theory). See Morrison (2012) for a detailed discussion of the philosophical implications concerning emergence in particular.

parameter thereby becomes an autonomous higher-level entity that defies a mechanistic description. In conclusion, I want to claim that theories of self-organizing systems, such as synergetics,—restricted to the justified descriptive reading of the slaving principle and thus refraining from taking the causal metaphors for the order parameter for real—explain the formation of system-wide patterns in terms of the endogenous interactions of the system's parts (hence *synergetics* for "working together"), and this fits nicely with the idea of mechanistic explanations.

13.4.2 Why Parts of a Mechanism don't need to be Spatial Parts

The second potential problem for a mechanistic reading of quantum laser theory is that the "parts" in the laser mechanism are not parts in the sense of spatiotemporal things. One source of this problem, which already applies to semiclassical laser theory, is that field modes are not individual things. For example, they can overlap and they cannot be traced through time. The other source of the problem is that we are dealing with quantum objects, which in general cannot be distinguished spatiotemporally.[17] Often, many quantum objects occupy the same spacetime region. Let us explore these potential problems a bit more closely.

One assumption in my above argument in favor of a mechanistic reading of laser theory is that field modes, or light quanta, are entities[18] that can feature as parts in a mechanism. But is it really sensible to understand modes of a field (classically possible states of oscillation) as parts? After all, different field modes can occupy the same region of spacetime. However, in the face of the wave-particle dualism, it seems just as legitimate or illegitimate to view light quanta as parts as it is to view, say, electrons as parts.[19] But this brings us to a more general point: What in general counts as a part in a mechanism? Rather than solving the problem of whether light quanta can be rated as parts, the reference to the wave-particle dualism shows that electrons and atoms may also be infected by the same problem.

---

[17] See Kuhlmann and Glennan (2014) for a more general and comprehensive discussion of whether quantum physics undermines the mechanistic program.

[18] As it is very common in ontology, I use the expression 'entity' as the most neutral ontological term, covering everything that exists from conventional things like dogs to properties and states-of-affairs. I only mention this because, in MDC's account of mechanisms, the term 'entities' is used more specifically in the sense of things or 'substances'.

[19] Falkenburg (2007, Chap. 6) explores the part-whole relation for quantum systems in more detail. She argues that the sum rules for conserved quantities such as mass-energy, charge, and spin are crucial for determining what we should rate as the constituents/parts of matter. On the basis of this criterion she draws a positive conclusion regarding the question of whether even the quanta of interaction fields such as the gluons in the quark model can feature as parts of quantum systems.

In the following I want to argue that it is a classical prejudice that *parts* of a concrete thing must always be spatially distinguishable entities.[20] In laser theory, field modes are sufficiently specified to function as independent *parts* that interact with the laser-active atoms. The field modes are not specified spatially, but with respect to their causal role. But that is enough for a mechanistic explanation to work. The decomposition of a compound system into components is a pragmatic matter that is ultimately justified by its explanatory success.[21] And in the case of the laser, understanding field modes as parts does the trick.[22] In many cases, it has no relevance where and even whether, say, objects O_1, O_2, and O_3 are located. What really matters is that, for example, (objects of type) O_1 is/are influenced by the behavior of (objects of type) O_3 in a specific way, while being unaffected by what (objects of type) O_2 does/do in the meantime. This situation is very common in complex systems research, where it is often only specified how the components are causally organized, whereas their spatial organization, if there is any, is left completely open.[23]

When we take field modes as parts of the laser mechanism, we stay very close to the mathematical treatment of lasers. Mathematically, field modes don't play any different role to laser-active atoms. Both are described by their own differential equations (which are coupled with each other). But this may be too much of a reification of field modes. Alternatively, it seems that one could stay with the conventional view and take the laser-active atoms as the crucial parts of the laser mechanism and the electromagnetic field as the interaction between the parts of the mechanism. In this case there would no longer be any need to relax the notion of parts by including entities that are not spatiotemporally distinguishable. However, I think it is nevertheless more appropriate also to treat field modes as parts of the laser mechanism. On the one hand, I argued above that the order parameter, i.e., the field mode $b(t)$ from above that "wins the battle"—because due to its comparatively long characteristic time scale for reaching its equilibrium value it can "enslave" the faster quantities—is no autonomous causal agent (see Sect. 13.4.1). On the other hand, the initial differential Eq. (13.3) apply to the whole spectrum of quantized field modes, which do real causal work. After all, "laser" is an acronym for "light amplification by *stimulated emission* of radiation", i.e., it is crucial for the emergence of laser light that the light field inside the laser cavity *causes* the atoms to emit radiation at a certain wavelength. And, I want to argue, it is most natural to treat those entities that do real causal work in a mechanism as *parts* of that

---

[20] I want to mention briefly that in current ontology there is a popular approach, namely trope ontology, which analyses things as bundles of copresent properties (understood as tropes, i.e., particularized properties). And many trope ontologists argue that properties should be seen as parts, although they can occupy, as constituents of one bundle, the same spacetime region.

[21] As an aside, Bechtel and Richardson (2010) distinction of decomposition *and* localization already implies that successful decomposition does not automatically lead to localized components.

[22] See Healey (2013) for similar considerations, but with a diverging aim.

[23] See Kuhlmann (2011) for detailed examples.

mechanism. Thus, in conclusion, I think it is more appropriate to rethink the notion of parts, and rate field modes as parts of a mechanism.

One last possible objection against field modes as parts is that their number is by no means constant, in contrast to the number of atoms. But this is not unusual in complex systems. We can clearly have mechanisms in complex systems where the parts can vary drastically. For instance, in convection cells of heated viscous fluids we can easily add and release molecules of the appropriate kind of liquid without changing or even stopping the workings of this self-organizing mechanism. Analogously, the changing number of field modes is no argument against rating them as parts.

13.4.3 Why Quantum Holism doesn't Undermine Mechanistic Reduction

The third potential obstacle for a mechanistic reading of quantum laser theory is that quantum holism may prevent us from decomposing the laser into different inter-acting parts, as is required for a mechanistic explanation. In general, the photons and atoms in a laser will be entangled with each other. Due to this entanglement, the subsystems (i.e., photons and atoms) are not in determinate states,[24] even if the whole laser is taken to be in a determinate state. Note that non-determinateness of properties differs from non-determinateness of states. In a sense, the latter is worse than the former. While the non-determinateness of properties can be dealt with in terms of dispositions or propensities, non-determinateness of states seems to pose a more serious threat to the applicability of the mechanistic conception in the quantum realm, because it may foreclose the ascription of properties to *distinct parts* of a compound system—no matter whether these properties are determinate (or 'categorical') or only probabilistically dispositional. To put it another way, I can't say everything relevant about one given quantum object without having to say something about other quantum objects, too, and this applies not just to their mutual spatiotemporal relation. This non-separability of quantum states is often called 'quantum holism'. Here we may have a strong form of emergence, because the reason why a given compound system (with entangled subsystems) is in a certain determinate or 'pure' state,[25] namely in this case a certain superposition, cannot be

---

[24] States comprise those properties that can change in time, like position, momentum, and spin (e.g., up or down for electrons). Besides these changing properties, there are permanent properties, such as mass, charge, and spin quantum number (e.g., electrons have the spin quantum number ½, which allows for two possible quantized measurement results, up or down, for any given spin direction).

[25] A pure state is represented by a vector in a Hilbert space. The contrast with a pure state is a mixed state, which can no longer be represented by a single vector. A mixed state can describe a probabilistic mixture of pure states.

explained in terms of determinate states of its subsystems.[26] In other words, the entangled parts of a compound system in a determinate state can no longer themselves be in determinate states. On this basis, Hüttemann (2005) argues that "synchronic microexplanations" do in fact fail in the realm of quantum physics, due to the notorious holism of entangled quantum systems.

Since the mechanistic conception of explanation is based on the reductionist idea that the behavior of compound systems can be explained in terms of their parts, it may look like the failure of reductionism due to the non-separability of quantum states could infect the mechanistic program, too. However, this is not the case because mechanistic explanations are concerned with the *dynamics* of compound systems and not with the question of whether the states of the subsystems determine the state of the compound system at a given time. In Hüttemann's terminology, the issue is diachronic and not synchronic microexplanations. As we have seen, in quantum mechanics the dynamics of a compound system is determined its by the Schrödinger equation—or the Heisenberg equation—where the crucial Hamiltonian that actually breathes life into the Schrödinger dynamics is the sum of all the "little Hamiltonians" for the system's parts and the interactions. Specifically, in quantum laser theory, in order to determine how the compound system evolves in time, all we need to know are the Hamiltonians for the subsystems, i.e., roughly the atoms, the light field, and the heat baths, and the Hamiltonians for their respective interactions. These Hamiltonians are simply added up. There are no tensor products for Hamiltonians and neither is there any entanglement of Hamiltonians.[27] In conclusion, one can say that, although quantum holism does mean that even the fullest knowledge about the parts of a given whole doesn't give us full knowledge about this whole, quantum holism does not undermine the mechanistic program of explaining the *dynamical* behavior of a compound system in terms of the interaction of its parts.

13.5 The Scope of Mechanistic Explanations

One could wonder now whether the requirements for something to be a mechanistic explanation are so general (abstract, loose) that practically any scientific explanation would count as mechanistic. Don't scientists always analyze complex phenomena, which are not yet understood, by reference to some kind of more basic items (call them 'parts') and then show how these items are related to one another (interact) to account for (bring about) the phenomenon in question? Well, 'Yes' and 'No'. The answer seems to be 'Yes', when scientists claim to have an explanation for some

---

[26] See Hüttemann (2005), who offers a very convincing study of the extent to which emergence occurs in QM, and correspondingly, 'microexplanations' fail vis-à-vis QM. Although Hüttemann's focus differs from that of the present investigation, his arguments are nevertheless relevant, with suitable adjustments.

[27] Note that this doesn't preclude the possibility of emergence in the sense of a failure of synchronic microexplanations.

dynamical phenomenon or law. In these cases they do in fact very often proceed in a mechanistic fashion. And one could even ponder the following claim: to the extent that science explains it does so mechanistically, and this fact is not undermined by QM. But this claim is arguably too strong. I don't want to claim that any scientific explanation is mechanistic, but rather that mechanistic explanations do not become impossible in the realm of QM, and are in fact widespread even there. So the answer is also 'No', since not every explanation or reasoning in science is mechanistic.

So why does quantum laser theory give us a mechanistic explanation? I think in this case one has to show in particular that the following non-trivial requirement is fulfilled: an account of how the components of the system interact in order to produce the phenomenon to be explained must lie at the core of the explanation. Note that this requirement for a mechanistic explanation is not fulfilled by the mere fact that an explanation makes reference to component parts and the way these parts are related to one another, as can be seen by looking at an example of a non-mechanistic explanation. In his famous derivation of black body radiation Planck (1901) calculates the entropy of a system of oscillators which he assumes to make up the walls of the cavity. This explanation refers to component parts, namely the atoms in the walls of the cavity, and to a certain extent it makes an assumption about how these parts are related to one another, but the interrelation of the constituent parts plays no important role in the explanation. Since it was already known in the nineteenth century that the spectral distribution of black body radiation is independent of the material and even the composition of the given body, one could to a certain extent assume just any kind of underlying processes in order to make the calculations as manageable as possible.

Nuclear physics is another context where mechanistic and non-mechanistic explanations coexist. Many explanations in nuclear physics are based on one of two very different models, namely the liquid drop model and the nuclear shell model. The liquid drop model treats the nucleus as an incompressible drop of nuclear fluid, and with this assumption it is possible, to a certain extent at least, to explain the energy as a consequence of its surface tension. Such an explanation is clearly not mechanistic. The nuclear shell model, on the other hand, describes the structure of a nucleus in terms of energy levels.

Another group of non-mechanistic explanations in physics concerns analyses that abstract completely from any processes that produce the phenomenon to be explained. In this group, I see for instance derivations and motivations based on conservation laws, symmetry considerations, and dimensional analysis.[28] One very simple example of the first kind is the calculation of the velocity of a falling object based on the transformation of potential into kinetic energy due to energy conservation, without any kinematical description whatsoever. Moreover, due to Noether's theorem, conservation laws are closely linked to invariances under

---

[28] Recently, Reutlinger (2014) has argued that renormalization group methods also yield non-causal explanations—and a fortiori non-mechanistic ones—not because of the irrelevance of micro-details, but because the mathematical operations involved are not meant to represent any causal relations.

certain symmetry transformations which can often be used in a very elegant way. Finally, a beautiful example of dimensional analysis is the derivation of the period of oscillation of a harmonic oscillator purely by considering the potentially relevant quantities and looking for a combination of these quantities that has the correct dimension—it turns out there is just one. [29]

Still another example of a non-mechanistic type of explanation is the derivation of special laws from more general laws in the covering law fashion. For instance, Kepler's laws for elliptic orbits of planetary motion can be explained using Newton's laws and certain approximations. In this case the two-body problem of the sun and a planet can be reduced to a one-body problem with a central force field around the center of mass of the two bodies. The reason this can be done is that the details of the interaction between the sun and the planet are totally irrelevant. And for this same reason, it should not be considered a mechanistic explanation. Yet another example of law-based non-mechanistic explanations are derivations based on thermodynamic laws like the ideal gas law. In these cases we do not refer to any causal mechanisms, but only state how certain macroscopic quantities are related to each other.

Finally, mechanistic explanations do not work for the simplest cases, such as the attraction of masses or charges in classical mechanics and electrodynamics, or the quantum harmonic oscillator and the behaviour of an electron in a magnetic field. This indicates that mechanistic explanations are not ruled out by the corresponding theory, but rather that some phenomena cannot be explained mechanistically because the system under consideration is either too simple or too fundamental. Thus, assuming that there is a bottom level in each theory, mechanistic explanations must come to an end somewhere, no matter whether we are dealing with quantum or classical physics.[30] Therefore, in this respect the main contrast is not classical

---

[29] See Sterrett (2010) for a philosophical analysis of the role of dimensional analysis in science.

[30] So can EPR style correlations also be explained by quantum mechanics? Imagine someone performs spin measurements on separated electron pairs that were emitted from a common source. Further imagine that our observer realizes that there are certain regularities in the results of two spin measurement devices. Each time she gets a spin up result in measurement device 1, she gets spin down in measurement device 2, and vice versa. Naturally, our observer assumes that there is a common cause for the correlations. By analogy, if you have pairs of gloves and each pair gets separated into two distant boxes, you always find a right glove in box 2, if you found a left glove in box 1. However, one finds that the electron pairs are correlated in a more intricate way: if you rotate the orientation of the spin measurement devices, you find the same kind of spin correlations again, even if you rotate by 90°. Since an electron cannot have a definite spin with respect to two mutually perpendicular orientations at the same time, the common cause explanation breaks down for the correlated spins of our electron pairs. In contrast, with quantum mechanics, it is possible to derive EPR style correlations from the basic axioms, namely from the unitary time evolution of states given by the Schrödinger equation and the resulting principle of superposition. But does this mean that EPR style correlations are explained? One could argue that in the framework of standard quantum mechanics, EPR style correlations are explained in a covering-law fashion. However, there is no explanation for why they come about, no causal story, and in particular no mechanistic story. Only particular interpretations or modifications of QM, such as Bohmian QM or the many worlds interpretation, may supply something like a mechanistic explanation.

mechanics versus quantum mechanics, but rather composite/organized systems vs. fundamental building blocks.

But the initial question may not yet be fully answered: Under which circumstances would laser theory not count as a supplying a mechanistic explanation of laser light? If one were to take Haken's quasi-metaphysical talk about the enslaving principle as an ontological commitment, then mechanistic explanation would become impossible to defend. While Haken has produced great achievements in laser theory, there is, as a consequence of Ockham's razor, no need to follow his metaphysical speculations, as we have seen in Sect. 13.4.3.

13.6 Conclusion

Mechanistic explanations are widespread in science, with the notion of 'mechanism' providing the foundation for what is deemed explanatory in many fields. Whether or not mechanistic explanations are (or can be) given does not depend on the science or the basic theory one is dealing with, but on the kind of object or system (or 'object system') one is studying and on the specific explanatory target. Accordingly, there are mechanistic explanations in classical mechanics, just as in quantum physics, and also non-mechanistic explanations in both of these fields. So not only are mechanistic explanations not corrupted by the non-classical peculiarities of quantum physics, but they actually constitute an important standard type of explanation even in the quantum realm.

References

Batterman, R.W.: The Devil in the Details. Oxford University Press, Oxford (2002)

Bakasov, A.A., Denardo, G.: Quantum corrections to semiclassical laser theory. Phys. Lett. A **167**, 37–48 (1992)

Bechtel, W., Abrahamsen, A.: Complex biological mechanisms: cyclic, oscillatory, and autonomous. In: Hooker, C.A. (ed.) Philosophy of Complex Systems. Handbook of the Philosophy of Science, vol. 10. Elsevier, New York (2011)

Bechtel, W., Richardson, R.C.: Discovering Complexity: Decomposition and Localization as Strategies in Scientific Research, 2nd edn. MIT Press/Bradford Books, Cambridge, MA (2010)

Bokulich, A.: Can classical structures explain quantum phenomena? British J. Philos. Sci. **59**, 217–235 (2008)

Cartwright, N.: How the Laws of Physics Lie. Clarendon Press, Oxford (1983)

Cartwright, N.: Nature's Capacities and Their Measurement. Clarendon Press, Oxford (1989)

Einstein, A.: "Zur Quantentheorie der Strahlung" (On the quantum theory of radiation). Physikalische Zeitschrift **18**, 121–128 (1917)

Falkenburg, B.: Particle Metaphysics: A Critical Account of Subatomic Reality. Springer, Berlin (2007)

Glennan, S.S.: Rethinking mechanistic explanation. Philos. Sci. **69**, S342–S353 (2002)

Haken, H.: Synergetics, an Introduction: Nonequilibrium Phase Transitions and Self-Organization in Physics, Chemistry, and Biology. Springer, New York (1983)

Haken, H.: Light: Laser Light Dynamics, vol. 2. North-Holland, Amsterdam (1985)

Healey, R.: Physical composition. Stud. Hist. Philos. Mod. Phys. **44**, 48–62 (2013)

Hillerbrand, R.: Explanation via micro-reduction. On the role of scale separation for quantitative modeling. In: Falkenburg, B., Morrison, M. (eds.) Why More Is Different. Philosophical Issues in Condensed Matter Physics and Complex Systems, Springer, Berlin (2015)

Hüttemann, A.: Explanation, emergence and quantum-entanglement. Philos. Sci. **72**, 114–127 (2005)

Illari, P.M., Williamson, J.: What is a mechanism? Thinking about mechanisms across the sciences. Eur. J. Philos. Sci. **2**, 119–135 (2012)

Kuhlmann, M.: Mechanisms in dynamically complex systems. In: McKay Illari, P., Russo, F., Williamson, J. (eds.) Causality in the Sciences, pp. 880–906. Oxford University Press, Oxford (2011)

Kuhlmann, M., Glennan, S.: On the relation between quantum mechanical and neo-mechanistic ontologies and explanatory strategies. Eur. J. Philos. Sci. **4**(3), 337–359 (2014)

Machamer, P., Darden, L., Craver, C.F.: Thinking about mechanisms. Philos. Sci. **67**, 1–25 (2000)

Morrison, M.: Causes and contexts: the foundations of laser theory. British J. Philos. Sci. **45**, 127–151 (1994)

Morrison, M.: Emergent physics and micro-ontology. Philos. Sci. **79**, 141–166 (2012)

Norton, J.: Approximation and idealization: why the difference matters. Philos. Sci. **79**, 207–232 (2012)

Planck, M.: Über das Gesetz der Energieverteilung im Normalspectrum. Ann. Phys. **4**, 553–563 (1901)

Reutlinger, A.: Why is there universal macro-behavior? Renormalization group explanation as non-causal explanation. Philos. Sci. (PSA 2012 Symposia) (2014)

Stephan, A.: Emergenz: Von der Unvorhersagbarkeit zur Selbstorganisation. Dresden University Press, Dresden und München (1999)

Sterrett, S.G.: Similarity and dimensional analysis. In: Gabbay, D., Paul Thagard, P., Woods, J. (eds.) Handbook of the Philosophy of Science, vol. 9. Elsevier, New York (2010)

Name Index

© Springer-Verlag Berlin Heidelberg 2015
B. Falkenburg and M. Morrison (eds.), *Why More Is Different*,
The Frontiers Collection, DOI 10.1007/978-3-662-43911-1

Titles in this Series

Quantum Mechanics and Gravity
By Mendel Sachs

Quantum-Classical Correspondence
Dynamical Quantization and the Classical Limit
By Dr. A. O. Bolivar

Knowledge and the World: Challenges Beyond the Science Wars
Ed. by M. Carrier, J. Roggenhofer, G. Küppers and P. Blanchard

Quantum-Classical Analogies
By Daniela Dragoman and Mircea Dragoman

Life — As a Matter of Fat
The Emerging Science of Lipidomics
By Ole G. Mouritsen

Quo Vadis Quantum Mechanics?
Ed. by Avshalom C. Elitzur, Shahar Dolev and Nancy Kolenda

Information and Its Role in Nature
By Juan G. Roederer

Extreme Events in Nature and Society
Ed. by Sergio Albeverio, Volker Jentsch and Holger Kantz

The Thermodynamic Machinery of Life
By Michal Kurzynski

Weak Links
The Universal Key to the Stability of Networks and Complex Systems
By Csermely Peter

The Emerging Physics of Consciousness
Ed. by Jack A. Tuszynski

© Springer-Verlag Berlin Heidelberg 2015
B. Falkenburg and M. Morrison (eds.), *Why More Is Different*,
The Frontiers Collection, DOI 10.1007/978-3-662-43911-1

The Language Phenomenon
Human Communication from Milliseconds to Millennia
Ed. by P.-M. Binder and K. Smith

The Dual Nature of Life
By Gennadiy Zhegunov

Natural Fabrications
By William Seager

Ultimate Horizons
By Helmut Satz

Physics, Nature and Society
By Joaquín Marro

Extraterrestrial Altruism
Ed. by Douglas A. Vakoch

The Beginning and the End
By Clément Vidal

A Brief History of String Theory
By Dean Rickles

Singularity Hypotheses
Ed. by Amnon H. Eden, James H. Moor, Johnny H. Søraker and Eric Steinhart

Why More Is Different
Philosophical Issues in Condensed Matter Physics and Complex Systems
Ed. by Brigitte Falkenburg and Margaret Morrison

Questioning the Foundations of Physics
Which of Our Fundamental Assumptions Are Wrong?
Ed. by Anthony Aguirre, Brendan Foster and Zeeya Merali

It From Bit or Bit From It?
On Physics and Information
Ed. by Anthony Aguirre, Brendan Foster and Zeeya Merali

Printed in the United States
By Bookmasters